"*Cooking G*                                        place in every environmentalists library. In simple language full of do'ts and don'ts for mindful cooking and eating, author Kate Heyhoe gives you all the information you need to shrink your cookprint, along with more than 50 recipes to get started. Not only does it thoroughly and thoughtfully present the new green basics of cooking, it provides the reasoning behind the recommendations, so as the climate changes, you can too, and so can your personal habits."

—**Linda Mason Hunter,**
author and pioneer in
America's green movement

"Let's save the planet one bite at a time! With *Cooking Green*, Kate Heyhoe gives us eaters the tools we need to preserve our natural resources while improving our dinner."

—**Sherri Brooks Vinton,**
author of *The Real Food Revival*

"I founded a children's cooking school 20 years ago and am thrilled to incorporate the new term cookprint into all of our classes thanks to Kate's vision and knowledge. With this clever and resourceful cookbook we can teach thousands of kids (and their parents) new ways to keep their bodies, minds, and their world a safe and healthy place to learn and live!"

—**Barbara Beery,**
children's cooking expert
ng cookbook author

**ALSO BY KATE HEYHOE:**

*Great Bar Food at Home*

*The Stubb's Bar-B-Q Cookbook*

*A World Atlas of Food*

*Macho Nachos*

*Harvesting the Dream: The Rags-to-Riches
Tale of the Sutter Home Winery*

*A Chicken in Every Pot: Global Recipes for the
World's Most Popular Bird*

*Cooking with Kids For Dummies*

# Cooking Green

*Reducing Your Carbon
Footprint in the Kitchen—
the New Green Basics Way*

*Hundreds of tips and over 50 energy- and
time-saving recipes to shrink your "cookprint"*

## Kate Heyhoe

Da Capo
∞
LIFE
LONG

A MEMBER OF THE PERSEUS BOOKS GROUP

Set in 10.5 point Warnock Pro by the Perseus Books Group

Illustrations © Nael Nabil/iStockphoto

Library of Congress Cataloging-in-Publication Data

Heyhoe, Kate.
Cooking green : reducing your carbon footprint in the kitchen : the new green basics way / Kate Heyhoe. — 1st Da Capo Press ed.
p. cm.
Includes index.
ISBN 978-0-7382-1230-2 (alk. paper)
1. Kitchens—Planning—Environmental aspects—United States. 2. Cookery—Energy conservation—United States. 3. Green movement—United States. I. Title.
TX653.H49 2009
643'.3—dc22
2008044602

First Da Capo Press edition 2009

Published by Da Capo Press
A Member of the Perseus Books Group
www.dacapopress.com

Da Capo Press books are available at special discounts for bulk purchases in the U.S. by corporations, institutions, and other organizations. For more information, please contact the Special Markets Department at the Perseus Books Group, 2300 Chestnut Street, Suite 200, Philadelphia, PA, 19103, or call (800) 810-4145, ext. 5000, or e-mail special.markets@perseusbooks.com.

10 9 8 7 6 5 4 3 2 1

FOR TW:

You get me energized!
Thanks for lighting up my life, every day, in every way.

*xoxoxox*

*Kate*

# Contents

# Introduction

*Did you know . . . 12 percent of total greenhouse gas emissions (or 14,160 pounds of carbon dioxide per household) result from just growing, preparing, and shipping our food.*

For a few years now, my green radar has been telling me that buying local and organic isn't enough. I suspected that, as a cook, I could do more to combat climate change. Lots more. The result: a treasury of practices that are as simple to integrate into daily life as changing light bulbs, yet focus on every step of the eating process. Some methods are old, some new, some I tweaked, and all are devised with good green benefits in mind. *Collectively, they're a whole new approach to cooking the basics.* These "New Green Basics" push the concept of "green cooking" far beyond the scope of just local, organic foods.

*How* you cook is as important as what you cook. Without abandoning your favorite recipes, you can bake, roast, broil, grill, and fry in vastly greener ways, saving fossil fuels, reducing greenhouse gases, and shrinking your *cookprint.*

"Cooking" has been a seriously underreported (yet substantial) greenhouse gas creator. In my book, it's the biggest way for kitchen-conscious consumers to take greener action. Shopping and cleaning tips tackle the sister-cycles of feeding activities. And these tips don't just help the planet. Many of my methods save time and money, too, yielding some unexpected side-benefits even for the cook.

To get the most bang out of your energy buck, why *not* start in the kitchen? After all, *appliances account for 30 percent* of our household energy use, and the biggest guzzlers are in the kitchen. After buying appliances with Energy Star labels, take the next big steps in the ways you use them.

*How* you cook directly relates to more efficient fuel use, and the less fuel used, the fewer greenhouse gases.

Plus, a single-family home spews more than *twice as many greenhouse gas emissions into the atmosphere as the standard sedan*—mostly from heating and cooling. Cooking can make a noticeable impact on household temperatures and how we adjust our thermostats. Anyone who's sweltered in a hot kitchen in summer knows the impact cooking has on local warming, not to mention global warming. A hot oven in winter can boost the room temperature, giving the household's central heater a break, but there are more efficient ways to cook.

The message: without changing your politics, or completely disrupting your routine, you can reduce greenhouse gases simply by rethinking what you must do every day: consume food. And with this book's green-method recipes and hundreds of idea-inspiring tips, you'll soon be serving your favorite dishes in new and greener ways, without thinking about the changes you've made to your daily routine.

As you will discover, the kitchen is ripe with opportunities for going greener. It's *the* place where you can make real choices and take direct control of your impact—without letting your family feel deprived, hungry, or stressed. In fact, everyone will feel better just knowing they're helping the planet—and they can do it one bite at a time.

## An Inspiring Story

Powerful ideas will migrate, if you share them.

George Marshall, an environmental campaigner in England, and his American-born wife, Annie, don't know it, but they were the inspiration for this book. "We have two young children, Elsa and Ned," writes George on their Web site. "We believe passionately that the world will be in a terrible mess for Elsa and Ned, and their children, unless there are dramatic changes in the way that everyone in rich countries lives. At the same time, we didn't want to throw out all the benefits of living in the early twenty-first century— appliances, central heating, artificial lighting—all bring a lot of freedom to our lives. The challenge was to keep some of these good things whilst very significantly reducing our impacts."

So George and Elsa set about redesigning an old yellow house, an ordinary English terrace house from the 1930s, with energy efficiency in mind. Their goal was to reduce the house's consumption of electricity, gas, and water by two-thirds compared with its average over the previous four years. In the first year, they met that goal with water consumption and halved their energy use. They kept working at it with further success, and today you can read all the nitty-gritty details of their adventure at www.theyellowhouse.org.uk.

The Marshalls launched a six-stage process of eco-redesign, ranging from thermal zoning their rooms to increased self-sufficiency of fuels and water. But what inspired me most were the ways they changed personal cooking habits to slash their carbon footprint (their *cookprint*).

I had an "a-ha" moment the instant I read George's account of "sit-boiling" and "haybox cooking." At last, someone who thinks like I do! I call it passive cooking, though the principle is the same: cook the same foods you always do, but with energy-efficient strategies to reduce fuel and water consumption. Earlier civilizations have employed many of these same tactics, but until the present century, rich countries hadn't felt the need to conserve. Now, with a scarcity of resources and a planet in peril, we most certainly do.

What the Marshalls and I share is motivation: we want to show people how to become more energy-efficient. The Marshalls tackle the issue from the ground up, literally, with building materials and home environment in mind. My focus is entirely on the building blocks of cooking: from farm and field, to fridge and fork, and cooking pan to kitchen fan.

Other like-minded cooks have, without knowing it, contributed to the strategies in this book. Admittedly, I am a cook first and scientist last, held together with a journalist's glue. Harold McGee, Robert Wolke, Alton Brown, and Shirley Corriher brilliantly explain the physics of cooking. Anyone passionate about the hows and whys of cooking should dive head-first into their books. The cooking chapters in this book show how to put fuel-efficient principles into practice, but the more details you understand, the better you'll be able to improvise with your own new green basics.

Powerful ideas migrate. Got your own ideas on how to shrink our cookprints? Share them at www.NewGreenBasics.com.

# The "New Green Basics" Behind Cooking Green

It's hard to believe that cooking with olive oil, ginger, and salsa was, at one time, novel and edgy. In the early 1980s, goat cheese was an Alice Waters–Berkeley type of thing, pizza came from a box or a chain, mayonnaise was mandatory in tuna salads, and wonton skins were exotic Chinatown ingredients. Pork was always cooked until overdone, and pasta sauces of raw tomatoes splashed across magazine pages as innovative entertaining. Organics came from health food stores, or if you lived in Austin, from a little neighborhood place known as Whole Foods Market.

Flip through cookbooks published prior to 1990 and you'll see why *The New Basics Cookbook* rocked. It was as heavy as the iconic classic *Joy of Cooking*, but in *The New Basics Cookbook*, Julee Rosso and Sheila Lukins codified a nationwide food revolution, and the country literally ate it up. Both cookbooks covered everything from meatloaf to mashed potatoes and abalone to veal. But with items like chèvre, Mexican ratatouille, and fig-and-pine-nut pizza, *The New Basics Cookbook* filled a widening gap between old-school Fannie Farmer traditional and what was settling into as freer, more expansive lifestyles, with cooks well versed in global ingredients. Today's generations are more likely to be stumped by Oxtail Stew than they are by Wakame Salad.

*The New Basics Cookbook* changed the way I and an army of others cooked, ate, and entertained, and it's still a model reference for cooks today.

Segue to "New *Green* Basics": everyday living with the environment in mind . . .

Inspired by the Lukins-Rosso book, *Cooking Green* organizes and outlines the *next* new basics of cooking: where greener methods aren't just options, they're the natural way of cooking for today and the future. It's the first practical guide for transforming old-school basics into eco-friendly cooking. It contains a manageable collection of recipes, designed to illustrate the concepts and put them into practice. Use them as they are, then apply the strategies to your personal repertoire. Each recipe comes with a handy Green Meter to highlight the way it saves fuel and water, and how it converts conventional methods to planet-friendly cooking.

# Before You Get to the Recipes...

In this book, Chapter 1 begins at the beginning: how much do you already know about fuel, water, and energy consumption? It tackles the fundamentals of how cooking works, and why. Refrigerators, cooking appliances (big and small), barbecuing, and the kitchen itself go under the spotlight in Chapter 2. Hands-on tactics for cooking green, with detailed methods and tips, flow through the middle of the book, from blue-oven cooking to green flame strategies on top of the stove. Does cookware make a difference? You bet it does, and Chapter 6 delivers the lowdown on fuel-efficient cookware.

Shrinking your cookprint includes the whole journey of food from farm to fridge to fork, from local pick-up truck to eighteen-wheelers with freezer containers. Chapters 7 and 8 work together: they outline ways to shop smarter, eat greener, and consume less. They dive into which foods are better for the planet, deciphering food labels, choosing sustainable seafood, and greening your plate with meat or without, and they introduce an array of fuel-efficient ingredients that need little or no cooking, like Vietnamese rice-paper wrappers.

With food comes waste. You really can control how much excess food and packaging you create, and Chapter 9 shows you how to scale back. (Not all waste comes from your kitchen: the restaurants you pick make a difference, too.) Be sure to check out the simple tips in Chapter 9 for making fresh foods last longer by storing them at the right temperature and in the proper containers; you'll waste less food and make fewer gas-guzzling trips to the store.

Good news: the recipes and tips don't end on these pages. NewGreen Basics.com updates basic strategies for greener lifestyles, from food to pets to kids. It's the brain trust behind this book. You won't find recipes for veal at NewGreenBasics.com, but if you're curious about Wakame Salad, dial up the site and grab a fork. (You'll also find thousands of general, international, and vegetarian recipes at www.GlobalGourmet.com, which we launched in 1994 as the Internet's first food and cooking site.)

Basically speaking, the concept of "new green basics" means choosing planet-friendly options in every aspect of our lives. Yes, it's about how to live greener, but it's also simply just another way to live, period.

*Kate Heyhoe*

# ONE

# A Cookprint, an Ecovore, and a Scientist Walk into a Kitchen …

What do you call the impact you make on the planet when you *cook*?

It's your *cookprint*—the entire chain of resources used to prepare meals, and the waste produced in the process.

## Defining Your Cookprint

A cookprint starts with food, in your garden or at the farm; it travels to your kitchen and continues in your fridge, freezer, or pantry. The cookprint grows larger every time heat or fuel is added, from a cooktop, oven, or small appliance. Discarded waste, whether it's organic produce trimmings, plastic packaging, or water down the drain, further colors the cookprint. So do the implements you cook with, the way you store leftovers, and how you clean up after the meal.

In short, the cookprint measures every meal's entire environmental impact. It's the total amount of energy and resources (from farm to fuel to fork) used in creating a meal. And it puts the cook squarely in charge of just how big, or how green, that cookprint will be—in ways that include but go far beyond buying organic or local, or eating meat or not.

# Why "Cookprint"?

In writing this book, I couldn't find the perfect term I wanted, so I created *cookprint* to define the serious but underreported impact of food and cooking on the environment, and how cooks, especially, can change it.

The *cook* in *cookprint* is a word of action. Just think of all the decisions, and all the physical steps, that go into answering the age-old question, "What's for dinner?" Even if you don't cook, *someone* cooks what you eat, and that contributes to your personal cookprint. *Carbon footprint* as a term measures carbon dioxide and other greenhouse gas emissions, but it doesn't reflect the impact of water usage or focus on the traceable path of food from earth to table.

Shrinking your cookprint forms the foundation of this book. It means questioning the things we take for granted, and making greener choices with every meal. Start by focusing on ingredients—where they come from, how they're grown, and how they're packaged. But don't stop there: consider *how* you cook your food, the type of energy, the amount of fuel consumed, the amount of water you use—and the amount of fuel and water you *waste*. In places and times where fuel is scarce, people never take fuel consumption for granted. Neither should we. Does that mean giving up slow-roasted foods or big, boiling pots of pasta? Absolutely not! But there are plenty of ways to stretch the fuels we use, every time we turn on the oven or fire up the burner, just by tweaking the recipes or methods you've always relied upon. Rethinking recipes and cooking methods is pivotal to greener cookprints.

As this book shows, a cookprint covers even the smallest details. It's about storing food in ways that use less energy, without sacrificing nutrition or flavor. Like making the refrigerator you already own more energy-efficient, and keeping fruits and vegetables fresh longer (for fewer shopping trips and less spoilage). Saving leftovers in glass containers rather than plastic ones or zipper bags, frying with energy-efficient skillets, and hundreds of other tips are included in the pages that follow.

## Take the Cook's Challenge: Can You Pass the Eco-Test?

If you really want to wire into a greener lifestyle, take this quiz and consider these issues. Do you know which are the greener options; that is, the ones that produce fewer greenhouse gases, use fewer natural resources, or come with a smaller cookprint?

1. Should the pilot light on a gas oven or water heater burn blue or yellow?

2. The refrigerator is the kitchen's biggest energy hog. What's the optimum temperature for your refrigerator?

3. Is it better to thaw frozen food in the microwave, in the fridge, or at room temperature?

4. If you're an eco-friendly cook, should your next cutting board be made of Corian, glass, bamboo, acrylic, or maple?

5. Which of these dried grains requires the least energy to cook at home: basmati rice, medium-grind bulgur wheat, or quinoa?

6. When it comes to energy-efficient cooking, rank these Asian noodles from least to most cooking fuel used: soba, rice sticks, bean thread, and udon.

7. Which is greener: a charcoal grill, electric grill, or a gas grill?

8. Does a convection oven produce more or fewer greenhouse gases than a traditional oven?

9. Which of these dried legumes cooks the quickest: chickpeas, limas, lentils, or black-eyed peas?

10. According to the U.S. Department of Agriculture (USDA), the term *organic,* with an official USDA Organic logo, means what percentage of the ingredients are actually organic: 100%, 95%, 70%, or 50%?

11. Do you save more energy if you run your dishwasher at midnight, noon, or 5:00 P.M.?

12. If your garbage disposal breaks, should you (a) call the repair service, (b) replace it with an Energy Star model, or (c) remove it and do without?

Answers on next page.

**1. Should the pilot light on a gas oven or water heater burn blue or yellow?** Think cool and blue. A yellow flame indicates a hotter temperature, consuming more fuel than necessary to ignite a burner. Better yet, opt for electric ignitions and ditch the pilot light completely.

**2. The refrigerator is the kitchen's biggest energy hog. What's the optimum temperature for your refrigerator?** In general, 37 to 40 degrees cools sufficiently without wasting electricity. Some foods last longer if kept in the colder sections of the fridge. Store them along the freezer wall (in a side-by-side) or in the back of the fridge; and never in the door. Or spot-chill them with freezer packs. For longer lasting freshness (and fewer grocery trips), keep dairy products at 33 to 38 degrees, meats between 31 and 36 degrees, and eggs at 33 to 37 degrees. Store fresh vegetables and ripe fresh fruits at 35 to 40 degrees.

**3. Is it better to thaw frozen food in the microwave, in the fridge, or at room temperature?** Thawing overnight in the fridge consumes the least amount of energy *safely*. Thawing at room temperature risks food contamination, and microwave ovens expend unnecessary power and thaw foods unevenly.

**4. If you're an eco-friendly cook, should your next cutting board be made of Corian, glass, bamboo, acrylic, or maple?** Glass is more eco-friendly than the man-made materials of Corian and acrylic, but chopping on glass is hard on knives. Bamboo is the most sustainable material, even more so than maple. It's strong, hard, and resists bacteria better than wood. Bamboo's downside lies in its traveling cookprint: bamboo comes from China. Maple hails from North America, but some forests are facing environmental stress. The best answer is to dig deeper: check into new cutting boards made of recycled cardboard, plastic, and cork; they perform well and repurpose materials that would otherwise go to waste.

**5. Which of these grains requires the least energy to cook at home: basmati rice, medium-grind bulgur wheat, or quinoa?** Bulgur wheat is partially cooked during processing, so at home it cooks in minutes, just by pouring hot water over it and letting it soak. Both rice and quinoa need to cook in boiling water for several minutes, so they consume more energy at home.

**6. When it comes to energy-efficient cooking, rank these Asian noodles from least to most cooking fuel used: soba, rice sticks, bean thread,**

**and udon.** Bean thread noodles (from soybeans) and rice sticks (from rice flour) soften completely just by soaking, so they take the least energy. Soba noodles (thin buckwheat noodles) and udon (thick wheat noodles) are boiled, though soba cooks faster. Better yet, the new green method for boiled noodles (including spaghetti) cooks them with half the fuel or less. (Chapter 8 explains how.)

7. **Which is greener: a charcoal grill, electric grill, or a gas grill?** Charcoal is the traditionalist's favorite, but propane and electric grills burn cleaner and make the better choices. Propane gas and electric grills still consume natural resources and release toxins, but charcoal releases more carbon monoxide, soot, and particles. Coupled with lighter fluid, charcoal cookouts also contribute more to ground-level ozone, especially in hot weather. (See page 50 for greener grilling, charcoal chimneys, and low-impact techniques.)

8. **Does a convection oven produce more or fewer greenhouse gases than a traditional oven?** A convection oven cooks 25–30 percent quicker than a traditional oven, so it produces fewer greenhouse gases. The type of fuel is the same, as are the type of emissions; there are just fewer of them with convection.

9. **Which of these dried legumes cooks the quickest: chickpeas, limas, lentils, or black-eyed peas?** Depending on type and age, dried lentils cook in just 10–30 minutes of boiling (without presoaking), making them the most energy-efficient of these legumes. To cook lentils with even *less* fuel, follow the "green flame" passive boiling technique on page 68.

10. **According to the USDA, the term *organic*, with an official USDA Organic logo, means what percentage of the ingredients are actually organic: 100%, 95%, 70%, or 50%?** Give yourself a green star if you picked 95 percent. Here's the breakdown: Products labeled as "100 percent organic" must contain only organically produced ingredients. To be labeled simply as "organic," 95 percent of the ingredients must be organically grown; the remaining 5 percent must come from nonorganic ingredients that are approved on a national list. Products billed as "made with organic ingredients" must be made with at least 70 percent organic ingredients, three of which must be listed on the back of the package, and the remaining 30 percent of the nonorganic ingredients must be approved on the national list. All three of these categories may display a certifying agent's logo, and only the last

one may not wear the USDA Organic label. (Confused? Flip to page 113 for more on labels, logos, and claims to watch for.)

**11. Do you save more energy if you run your dishwasher at midnight, noon, or 5:00 p.m.?** Because electricity at power plants is generated more efficiently during off-peak hours, midnight saves fuel at the source.

**12. If your garbage disposal breaks, should you (a) call the repair service, (b) replace it with an Energy Star model, or (c) remove it and do without?** Lose it, don't use it. Garbage disposals bring unnecessary energy and water consumption to the waste process. Composting, and even regular trash disposal, are better options. Garbage disposals don't come with Energy Star ratings.

*The Eco-Cooking Quiz*: *Score yourself Light Green if you got six answers correct, Solid Green if you answered nine questions correctly, and Bright Green if you know—and practice—them all.*

## Boiling Down Your Cooking Efficiency

If Jed Clampett's oil wells were as efficient as our cooking stoves, he'd still be a poor hillbilly in the Ozarks, not a rich hillbilly in Beverly Hills.

Imagine that, like Jed, you strike oil on your property, and you set up four pipelines to capture it. But all your pipes have rusted holes in them. For every 100 gallons of oil entering a pipeline:

- the red pipe spills 93 gallons
- the blue pipe spills 60 gallons
- the yellow pipe spills 86 gallons (and maybe more)
- the purple pipe spills 26 gallons (and maybe more)

You'd be wasting a whole lot of oil. And so do our kitchens. As fuel-efficiency goes, the red line's a gas oven. The blue line's a gas cooktop. The yellow line's an electric oven, and the purple pipe represents an electric cooktop (both of which waste even more energy if their electricity is generated from fossil fuels instead of renewable sources).

*(continues)*

## Thermodynamics in Three Minutes or Less; or, What Makes Cooking So Complicated?

*Heat is energy. It's everywhere and it is always on the move, flowing out as it flows in. It roils the chemical innards of things, excites their molecules to vibrate and crash into each other. When we add a lot of heat energy to foods, it agitates those innards enough to mix them up, destroy structures and create new ones. In doing so it transforms both flavor and texture.*

—Harold McGee, New York Times, *January 2, 2008*

As you read this book, you'll run across a few terms and processes that are handy to know, if not in detail, at least in principle.

---

(Boiling Down Your Cooking Efficiency, *continued*)

The U.S. Department of Energy's snapshot of cooking efficiency skips the 1960s TV analogy, but the raw data's the same and looks like this:

*Energy Efficiency*

| (amount of energy reaching the food) | Appliance |
|---|---|
| 6% | Gas oven—standard |
| 7% | Gas oven—self-cleaning |
| 40% | Gas cooktop |
| 12% | Electric oven—standard* |
| 14% | Electric oven—self-cleaning* |
| 74% | Electric cooktop* |

* *Electricity Note:* These percentages reflect the end-use appliance efficiency. If not using renewable energy, the energy loss in converting fossil fuels to electricity significantly lowers efficiency, as explained on page 26, The Footprint's Shadow.

# Heat Capacity

Heat capacity is the amount of heat required to raise a medium by 1 degree C. Water's heat capacity is about four times greater than that of air. In other words, it's much harder to change the temperature of a given amount of water, and requires more fuel, than it does to change the temperature of the same amount of air.

Not only that, but water heats up, and cools down, more slowly than air. This logically means that air cools faster than water.

All of this might suggest that oven cooking is more efficient than boiling water on a stove. Wrong. You also need to consider such pesky issues as:

- how heat is transferred (by conduction, convection, or radiation, or a combination)

- the nature of liquids and gases (water's maximum temperature is its boiling point: 212 degrees F, or 100 degrees C; most oven cooking takes temperatures into the 300–500 degrees F range.)

- the material of the vessels used

And then there's the physical structure of the oven and the cooktop themselves, and the nature of how each performs.

## Some Interesting Things About Heating Water and Air

- It takes seven times as much energy for water to jump from just below the boiling point to actual boiling. So, if very hot but not boiling water will get the job done, then you're saving fuel.

- You can keep adding heat to water, but once it reaches the boiling point, water will never become hotter, no matter how much fuel you use.

- Water is the same temperature at a rolling boil as it is at a gentle boil. Vigorous boiling may churn foods around more, but it's not any hotter than a regular boil. It just burns more fuel.

- If you stick your hand into a boiling pot of water, you'll burn yourself badly—and instantly (so don't do it). But you can reach into an oven that's twice as hot and not burn your hand. Likewise, you can pass your finger through a candle flame, and it won't burn. But touch a metal pot sitting over that same flame, and you'll be reaching for the first aid kit. Why? Because air has a lower heat capacity than water or metal.

- Heat flows from hot to cold. That's why in a pot of steam, for instance, cooking occurs when hot vapor gloms onto cold food. It's also the mechanism behind "carry over" cooking, in which the hot outer area of a roast will continue to cook the interior, even without additional heat.

## Understanding Heat

Heat = energy. Heat is transferred by three different types of processes. Cooking is a function of temperature and the rate of heat transfer. But rarely does a food cook by just one process.

Conduction heat transfers by direct contact. That is, atoms and molecules within a material collide (come in contact with each other) to create kinetic energy, or heat. Generally speaking, the tighter the atoms and molecules are to each other, the more vibrations they create, and the more heat is transferred. Some materials conduct heat better than others. Liquids and gases (like water and air, respectively) have lots of distance between their molecules, making them poor conductors of heat. Metals are good, fast conductors of heat. Glass and ceramics are better *insulators*, that is, they conduct heat but they do it slowly. Solid foods also behave more like insulators, slowly letting the heat penetrate from the outside to their interiors. All types of materials (from foods to pans) conduct heat differently, and some are more efficient than others. "Thermal conductivity" refers to a substance's ability to take heat through conduction. Highly conductive materials transfer heat quickly, and materials of low conductivity transfer heat slowly. But the ability to move heat rapidly doesn't automatically create rapid temperature change. You'll learn more about these concepts—and how to put them into practice—in the cookware chapter (Chapter 6).

Convection heat transfers by movement. Molecules actually move around, not just vibrate. They move from a warm substance to a colder one. When they move, they bring their heat with them. Ocean currents and winds are examples of convection. In cooking, liquids and air act in the same way. Have you ever heard that hot air rises? The same thing happens in an oven, and the movement of hot air within the confined space creates currents. In a pot of water, heat creates currents that lead to boiling. When steaming foods, water converts to steam, and the steam moves in currents within the enclosed pot. In a convection oven, a fan circulates the heat into forced currents, resulting in more even, and more energy-efficient, cooking. (Breeze through Chapter 3 for the section on convection ovens.)

Radiation heat transfers by electromagnetic waves (not the same thing as nuclear radiation). More accurately, thermal radiation happens when atoms and molecules release energy they've absorbed in the form of infrared waves. When the waves hit a food's atoms, they spark movement within those atoms. They work without a medium (like a pan or liquid) to come between the food and the heat source. The energy literally beams into the food. Microwave ovens, for instance, agitate water molecules. All things emit radiation constantly, but mostly at levels so low they're inconsequen-

## Popcorn 101: Cooking Basics

Popcorn is an easy way to understand the mechanics of conduction, convection, and radiation:

**Conduction:** Heat a layer of oil in a pan, add the popcorn, and cover the pot. The heat transfers through the pot to the oil, and from the oil to the popcorn. Pop!

**Convection:** Fire up a hot-air popcorn popper. Heat moves through the chamber in currents, and as this hot air transfers to the kernels, they absorb the heat and burst.

**Radiation:** Microwave a bag of popcorn, or microwave kernels in a glass popping container. The microwaves agitate the molecules of the popcorn kernels, creating heat from within each kernel, until they pop—without the need for swirling hot air in the chamber or transfer of heat from a pan.

Admittedly, these boiled-down fundamentals of heat and cooking barely skim the surface of the science, but are adequate for the pages ahead. Naturally, all rules have exceptions when certain conditions are applied—for example, using a pressure cooker to change the boiling point of water (we'll tackle that steamy issue in Chapter 6).

tial. When cooking at higher temperatures, like those of a barbecue grill, broiler, or very hot oven, infrared radiation generates waves that are potent enough to cook foods; however, at lower temperatures, conduction and/or convection methods are more efficient ways to cook. (The spectrum of radiation isn't just about cooking. Starting at the lower end of the scale, your car radio emits weak waves; and higher up, microwave towers relay cell phone calls, and microwave ovens heat frozen dinners. On the upper end of the spectrum, ultraviolet waves cause sunburns, and X-rays show us our bones.) Not all materials absorb radiation in the same way, and radiation waves themselves differ in their capacity to generate heat. (For more details on the infinite nuances of radiation, and other mysteries of kitchen science, dial up Harold McGee's www.curiouscook.com.)

## The Ecovore's Dilemma

"Eat food, not too much, mostly plants."

So goes Michael Pollan's healthy eating mantra. Pollan is the bestselling author of *The Omnivore's Dilemma* and *In Defense of Food.* His sound bite sounds simple, but his books wrestle with the complexities of how we feed ourselves—weighing in on nutrition, politics, agriculture, environment, and ethics. Even he admits the solutions for healthy eating aren't quite as straightforward as the mantra suggests (but it's a good tool for encapsulating the bigger message).

Eat food, not too much, mostly plants. As an *ecological* directive, it's also good advice, just as eating organically and locally are. But as climate change escalates, any list of do's and don'ts—like last year's computer—becomes outpaced even before it's released.

Which is why I've cobbled together a more fluid green meme: "ecovore." *An ecovore eats foods that are raised and grown in harmony with the environment, currently and for the foreseeable future, locally and globally.* But like Pollan's mantra, it's not as simple as it sounds. What those foods are at any given time, or in any given place, is in constant flux, because of changes in ecosystems, economics, and technology. And with global warming, the ecovore's challenges are just beginning.

The impact on the food chain appears particularly alarming. One of the most extensive studies, a 2008 report by the U.S. Climate Change Science

Program, concludes that climate change will continue to profoundly impact our land and water resources, agriculture, and biodiversity. Higher temperatures and decreased or erratic precipitation will spawn more crop failures (especially grain, oilseed—as in canola and other cooking oils—and horticulture like tomatoes, onions, and fruit). Livestock mortality will further increase while productivity decreases. Water will become an even bigger issue, especially in the West, with increased drought and forest fires, smaller mountain snowpack, and earlier snowmelt to drastically reduce drinking water.

Whenever natural resources change, crops change with them, and so does the list of do's and don'ts. If water is scarce, local farmers may switch from irrigated peach orchards to drought-resistant grains. Naturally, some seasons produce smaller yields, but when this happens over time, and farmers must alter what they grow, the ecovore's options for eating locally also shift.

In a world of climate change, "sustainability" changes rapidly. Where you live, and what you know about what's happening around you, matters.

Even those who've tried it admit that the "locovore" hundred-mile diet is not a realistic lifestyle for today's consumers. We thrive on flavors from around the globe, seasonal or not. Certainly, the long-term solution to a healthy and diversified diet (for both the people and the planet) starts with organic practices, but it also depends on reducing transportation emissions, and preserving food freshness and nutrition without ozone-depleting gases. With a global infrastructure based on carbon-free transportation, we could feed the world, sustain all the farms, preserve natural environments, support biodiversity, and support healthy economies. It's possible, not yet practical, and may never happen in full, but we need to raise our goals and realize that even the smallest steps matter. If transportation technology could deliver foods from another continent without harming the environment, then cookprints would automatically shrink. It's a worthwhile dream and just as wacky an idea as putting a man on the moon.

The global economics of food also color the ecovore's diet. If poorer nations are suffering because corn costs have skyrocketed due to demand as a biofuel, or if a global rice shortage is literally hitting below the belt, should you buy corn and rice? Instead of these grains, would opting for quinoa, even if it's grown in South America, be a wiser choice instead? An ecovore would ideally choose domestically grown quinoa, but until recent times, such an item did not exist.

Speaking of South America, it makes sense to grow things in places where they grow best, even with a transportation footprint. My morning breakfast would be very sad without shade-grown coffee from sustainable farms in the rainforest, and it would be a tougher life for those farmers as well. (By supporting rainforest goods, we in turn prevent the destruction of rainforests by loggers and land developers. So even though coffee beans create transportation carbon, growing them sustainably yields beneficial green impact.)

Being a vegan or vegetarian is not the same thing as being an ecovore. The definition of *ecovore* is inherently one of fluidity. It demands a series of judgment calls based on conditions at the time and place of purchase. This season's salmon may be sustainable, but next year it may not. And conversely, as we make progress, what casts a carbon footprint last week may not be an issue tomorrow.

Going greener is all about choices. This book is full of do's and don'ts for mindful eating. But it also works to provide the reasoning behind the recommendations, so as the climate changes, you can too, and so can your personal habits.

Without succumbing to eco-anxiety, we can start fixing what's broken by setting greener goals—person by person, aisle by aisle, and kitchen by kitchen. Eat food, not too much, mostly plants—and I would add, in this order: mostly local, mostly organic, and mostly fuel-and water-conserving. In a lifestyle of climate change, it's the new green basics of cooking.

# TWO

# Where You Cook: Greening Your Kitchen Zones

Many eco-consultants these days will, for a fee, analyze ways to green your kitchen. Like a "green consultant," this chapter walks you through each area of the kitchen, analyzing ways to optimize your refrigerator, make foods last longer, wash dishes with less water, scale back fuel use with small and large appliances, and clean with a green sheen. We also stroll outside to the grill, for tips on greener grilling.

The first rule: make do with what you've already got. Reduce, reuse, recycle, and *replace* only when it makes good green sense. If your kitchen is woefully inefficient, and you're planning on a whole or partial kitchen remodel, options for green materials expand daily. But be diligent in your research. Bamboo flooring, for example, is green and sustainable in one sense, but if it's grown in Asia and shipped across land and sea to get here, it dumps a hefty chunk of transportation carbon into your cookprint. Appliances carry huge manufacturing footprints. Replace inefficient energy hogs with Energy Star appliances, but seek out greener options before tossing them. If they're in good shape and run well, keep using them, or donate them to a place where they'll get a second life.

## The Kitchen Zones: Cold, Hot, Wet, Dry— and the Outdoor Zone

The biggest savings of fuel and water happen in three zones: the refrigerator/ freezer or "cold" zone; the stove, oven, cooktop, microwave, and other cooking areas known as the "hot" zone; and "wet" zone items—sinks and, for the purpose of this book, water heaters. "Dry" zones are workplace countertops, including overhead and task lighting. And since we've become a nation of sun-worshippers and grilling geeks, I'm designating outdoor barbecues and solar cooking as the fifth zone.

## The Cold Zone: Too Cold, or Not Cold Enough?

Let's start with the elephant in the room. When it comes to home energy consumption, refrigerators eat up 11 percent of the entire home's electricity (as much as all the lights combined). So use them as efficiently as possible, even if they're Energy Star–certified.

Don't turn up the thermostat on your refrigerator or freezer just to make things colder, but do keep in mind that when fresh foods are stored properly, at their specific optimal temperatures, they stay fresh longer, meaning fewer gas-guzzling trips to the store. Use some inexpensive refrigerator and freezer thermometers and check them seasonally; you'll likely need to adjust the thermostats every winter and summer.

### Greener by Degrees

In general, 37 to 40 degrees F cools sufficiently without wasting electricity, even though 35 to 38 degrees is a better range for extending the freshness of foods. Make the most of the cold spots in your fridge without turning the thermostat down: these are located along the freezer wall (in a side-by-side) or in the back of the fridge—never in the door. You'll get a few days' extra mileage by keeping dairy products and eggs at 33 degrees, meats at or just above 31 degrees, and most fruits and vegetables between 34 and 40 degrees (citrus fruits are best at 39 degrees).

### Freezer Packs Make Meats Last Longer

Fresh meat and poultry can last up to three days longer if stored at 31 degrees F, well below the standard fridge temperature. The spoilage rate slows down, without solid freezing. Some fridges have programmable bins with this setting, but check this out: to increase the chilling power of a standard meat bin, toss in one of those frozen blue-ice packs, the kind used in picnic coolers. Or, if you're planning on slow-thawing a frozen package of meat, do it in the meat bin. It will take a couple days to thaw and will drop the other meats to a lower temperature at the same time. By the way, fish markets have loads of freezer packs; ask for a pack or two to keep your fish cool on your way home and in your fridge. Wash the pack well with a little vinegar in the water to remove any odors, and reuse it whenever you need to chill.

### Seven Green Ways to Use a Freezer Pack

Freezer packs thaw slowly, especially inside a refrigerator, and they don't waste water like melting ice can. Keep some handy in the freezer, then use them:

- In a bowl instead of ice when shocking vegetables in "ice water"

- In the refrigerator meat bin (a lower temperature can extend freshness up to three days)

- To take up vacant fridge or freezer space (the motor won't need to work so hard)

- To keep groceries chilled in an ice chest (less pressure to rush home, so you can do more errands while you're out; plus perishables last longer when kept consistently cold)

- For ice chest–chilled drinks on patios and at barbecues (with fewer trips indoors, the house and fridge both stay cooler)

- To keep refrigerated fish at its peak of freshness

- Under your milk carton (dairy products prefer 33 degrees F, slightly cooler than most fridges)

## Get the Most Out of Your Fridge

For a more energy-efficient refrigerator:

- Keep the refrigerator coils clean by vacuuming them occasionally; they'll function more efficiently.

- Make sure the seal isn't worn out: close the door with a lit flashlight inside. If you can see light from the door seal, replace the seal.

- Let air circulate around the refrigerator exterior. Heat from the coils needs to escape; if it doesn't, the unit works harder. Leave a few inches of breathing room at top, especially, and around the sides. Avoid placing items that block circulation on top of the fridge.

### *If Being Green Means Needing to Buy a New Refrigerator*

- Compare the Energy Star labels to see which refrigerators are most efficient. As of 2008, refrigerators must exceed federal efficiency standards by 20 percent (up from 15 percent) to qualify for Energy Star certification.

- Opt for a freezer on the top. Side-by-side models use 10–15 percent more energy, and bottom freezer drawers fall in between the two.

## Energy Star: What Qualifies?

In 2007 the Energy Star program helped Americans prevent greenhouse gas emissions equivalent to the output from 27 million cars—and they saved $16 billion on utility bills. Energy Star–qualified appliances are designed to use 10–50 percent less energy and water than standard models, according to guidelines set by the Environmental Protection Agency (EPA) and U.S. Department of Energy.

When shopping for appliances, pay attention to the yellow Energy Guide label. It shows how much energy the appliance consumes, compares energy use of similar products, and lists approximate annual operating costs. Only the products that exceed efficiency standards in their category qualify for an Energy Star label. For instance, Energy Star refrigerators are at least 20 percent more efficient than the minimum federal standard.

Not all types of products can be Energy Star–certified. What qualifies? Refrigerators, freezers, dishwashers, washing machines, home heating and cooling products, televisions, certain building materials, light bulbs and fixtures, and some office equipment.

However, there's no Energy Star program for microwave ovens, clothes dryers, water heaters, ovens, stoves, cooktops, or small appliances like blenders and slow cookers. For most of these products, the U.S. Department of Energy offers consumer guides and tips online, and consumer advocacy resources like *Consumer Reports* frequently perform and update their own tests.

- Automatic ice makers and in-the-door dispensers (for cold water and ice) increase energy consumption by about 15 and 10 percent, respectively. If you don't need these features, don't buy them.

- One is better than two. A single large refrigerator is more efficient than two smaller ones. But the best choice is to be like Goldilocks: find the size that fits just right.

- Automatic defrost models are now as efficient as manual defrost models.

- Think outside the standard box: invest in a high-efficiency refrigerator, if you can afford it. It's expensive but comes with improvements to energy efficiency, insulation, compressor, door seal, temperature accuracy, and overall savings.

- Finally, fill 'er up: Keep your fridge well stocked (you'll use less energy to keep it cool). Fill up empty spaces and barren shelves with bottles of water (take them out whenever you need the space). Ditto for the freezer: use freezer packs and bags of ice to fill the voids.

## Soaking Up the Wet Zone: Water Conservation in the Kitchen

Don't let good water go down the drain. Healthy water supplies may not be an issue where you live (now), but many cities, states, and homeowners face water rationing at least some days of the week. Water conservation in the kitchen is a habit for everyone to adopt, and it's easy once it's on your radar.

### *Dishwashers for Daily Use*

Most of today's machine dishwashers consume 1–2 kilowatts of electricity per load, and 6–14 gallons of water, depending on the type of setting. Despite their carbon footprint from manufacturing and transportation, dishwasher machines can be more efficient than washing dishes by hand, if you make proper use of their efficiency.

You can green up your daily dishwasher use if you:

- Scrape food off plates rather than rinsing.

- Run only a full load.

- Set to an air-dry cycle rather than heat dry.

- Wash with biodegradable detergents.

- Pick the right setting: light rinse, pot-scrubber, or whatever's most appropriate.

When buying a new dishwasher, opt for an Energy Star model and:

- Consider soil sensors: this feature automatically adjusts water according to how dirty the load is.

- Pick the right size: compact or standard capacity (too large a machine wastes fuel and water, and too small may require multiple loads).

- Install the unit away from the refrigerator: dishwasher heat will cause a fridge to work harder.

- Recycle your old unit: working energy-efficient units can be offered to charities like Habitat for Humanity or Goodwill (contact your community recycling program for older units).

## Getting the Temperature Right

When it comes to hot water, washing by hand can't compare to the temperatures that machine dishwashers use. Both will rid your dishes of bacteria, but they do so in slightly different ways.

Most dishwashers start with normal household hot water, then bump it up to 140–145 degrees using an internal booster heater. From a health standpoint, this is a good thing, as it wipes out beaucoup bacteria. With a dishwasher's booster, there's no need for your household water heater to stay set at 140 degrees (usually the manufacturer's default setting). The human skin can tolerate water to 110 degrees, but only in short bursts of time. So with a temperature-boosting dishwasher, you can maintain overall household hot water at just 115–120 degrees, which is plenty adequate for handwashing and laundry. Set the water heater to between medium and low, and let the dishwasher raise only the water it needs.

### The Bacteria–Hot Water Relationship

Hand-washing dishes uses water about 30 degrees cooler than dishwasher machines, but it's the role of surfactants (soap) that get hand-washed dishes clean enough to eat on. Water that's hot but tolerable loosens food, grease,

## Power Down in Peak Hours

Run your dishwasher at night. By simply switching your energy consumption to off-peak hours, you can save fuel at the source. Power companies bring power plants online and offline during the day, according to demand. Large energy savings happen, though, when energy is used more steadily: fewer peaks and valleys improve the power plant's own efficiency. In most places, the hours between midnight and 4:00 A.M. are best, but even if you skip the hours between 4:00 P.M. and 8:00 P.M. you'll be avoiding the energy rush-hour. Besides dishwashers, anything else you can run in off-peak hours will save energy (like TIVO-ing *The Daily Show* at midnight, rather than during prime time).

and bacteria, while the soap lifts these elements off so they're washed away with the rinse water. Technically, 110-degree water doesn't kill bacteria, but combined with soap and scrubbing it does the job of removing the nasties and sending them down the drain. (Note that not all bacteria succumb to hot water alone. For some, it's a time-and-temperature whammy that does the trick.) And this doesn't apply just to dishes. Cleaning your hands with hot water, regular soap, and at least 20 seconds of rub-a-dub scrubbing will clean your hands without the need of antibacterial soaps.

### Saving Water at the Sink

Ever wonder if it makes more sense to clean items in a full sink of water, or simply rinse them under running water, as with produce, dishes, greasy pots, or hand-washable fabrics? Sometimes, rinsing an item under running water alone makes sense. Other times, washing items in a full sink of water, then rinsing them in clean water, is both the healthier and greener strategy. As drought conditions worsen (and to conserve water generally), it helps to understand how much water you're using, and options for using it more efficiently.

But first: *How much water does your sink hold?*

To find out, multiply the length, depth, and width of your sink (to a practical water level, not to the very top), for the total cubic inches. Then multiply cubic inches by .0043 to get the number of gallons of water your sink holds.

I have a split sink, with two basins. Each basin measures 7.5 x 15.5 x 16 inches when filled to about 2 inches below the top. These three dimensions multiplied equal 1,860 cubic inches, which when multiplied by .0043 yields 8 gallons (7.998 gallons, rounded up). So each basin, when usably full, holds about 8 gallons of water. (By comparison, my Energy Star dishwasher uses less than 6 gallons of water for a normal load, and about 7 gallons for a heavy load.)

How does this compare to running water? I plugged up the sink, and using the sink's aerated faucet with moderate pressure, I proceeded to "shower" rinse a head of romaine lettuce, draining the leaves into an over-the-sink colander. When done, I had collected under two inches of water, or approximately two gallons.

As an alternative method, I filled the sink half-full, then swished a head of lettuce leaves in the water, letting the grit settle for a few minutes. I needed four gallons, or half a sink's worth, of water to keep the lettuce above the grit level, compared to two gallons for shower-rinsing. If I was planning to recycle the swished water, like rinsing more vegetables in it, then swish-and-rinse makes more sense than the quick-shower method. And the best tactic would be the shower method, with the water captured in a tub and reused.

However you do it, capturing water prevents senseless water waste. One woman I know in a drought zone keeps a pitcher under her faucet to catch water before it goes down the drain, and doing so keeps her plants well hydrated. As long as the water isn't too gritty, it's fine for precleaning really messy pots or dishes, soaking labels off jars, or rinsing containers destined for recycling. If the water's not tainted, I often dump in some potatoes, scrub them, and then rinse under running water to remove surface dirt. Oranges and melons, too; bacteria on the skin can transfer to the flesh upon cutting, and citrus should always be washed before zesting. Just be sure the final water is clean and rinses away any loosened dirt or bacteria.

### Do Sink Those Gritty Edibles

The best way to wash really gritty produce, like spinach, is to swish the leaves around in a sink or tub of water, then let them rest for a bit. Gravity will pull the dirt to the bottom. Scoop out the leaves to drain, or if really gritty, repeat the process with clean water.

### Tubby Tip

A plastic tub under the faucet traps water, but if you don't want to tie up your sink, set a tub or basin near the sink, and fill it with water. Use the tub for the first wash, and the sink for a final shower-rinse.

### Hot Water Tip

We do it all the time: turn on the hot tap and wait for hot water to reach the faucet. What a waste! Capture this squandered water in jugs and use it for plants, washing the dog, the floors, the car, the bikes, or prerinsing spinach with it. This isn't grey water, but because it runs through the hot water heater and pipes, it may contain more minerals (copper, iron, lead, etc.) than water from the cold tap. For faucets with a single hot/cold control, always push the control to "cold" when turning off the water, and push it to "hot" only when you really need hot water.

## Faucet Aerators

Handy faucet aerators snap or screw right on, and the best models offer these features: swivel action, aerated jet flow, wide spray, and one-touch flow restrictor (to go from single stream to wide spray with just a tug). You'll save money, use as much as 50 percent less water, and have greater water control. Aerators work as well as handheld sprayers installed on many sinks.

## Water Heater Kitchen Tips

On-demand and solar water heaters make good green sense, but they're still outnumbered by conventional water heaters with barrel-style tanks. These conventional water heaters are a conservationist's triple whammy: they consume a hefty burst of energy to heat water, more standby fuel to keep it hot, and then lose a chunk of heat as the water travels through the pipes. So remember: whenever you turn on the hot water tap, you're burning fuel.

- If you don't need hot water, don't run it. This seems obvious, but there's more to it. Every time you turn on the hot water, a quart or more of water that was heated goes cold. The hot water valve draws water from the heater, and if it's not used immediately, this water in transit to the tap cools down as it sits in the pipes. So you may or may not be wasting water, but you are wasting the energy to heat it.

- Insulate the water heater and the pipes from the hot water heater to the kitchen, and bathrooms.

- Set the water heater temperature to 120 degrees F or below (medium to low setting).

- Drain the tank yearly, especially in areas with high mineral content. Sediment collects in the water heater, which can damage the unit over time. (Drain with a hose into your garden.)

- If you're going to be away for at least three days, drop the heater setting to as low as it can go, or turn it off.

## Disposals: Yes or No?

What to do with kitchen scraps? Some tactics are greener than others:

**Garbage disposal:** you don't need one. Garbage disposals eat up electricity and water and come with the occasional repair bill. They're only good for soft matter, not for bones and seafood shells. The chewed-up garbage either creates sludge in septic systems or travels to a wastewater plant, where the water is treated and processed. The solid matter is filtered out, then driven by truck to a landfill (while the liquid typically channels into waterways, polluting and damaging the bio-balance.) You're better off not using the garbage disposal at all. And if you never install one, you're also skipping the energy used in the manufacturing process, transportation, and installation of the unit itself.

**Kitchen trashcan:** scrape your plates (rather than rinsing them) before placing them in the dishwasher. It saves water, and it skips the path from garbage disposal to treatment plant and landfill.

**Green bins:** some municipal waste management programs offer "green bins," an excellent option for consumers who don't compost. Kitchen scraps as well as yard waste go into green bins, which are driven to composting sites. The compost goes to farmers and back into the soil.

**Composting:** compost is nature's way of recycling. Eventually, all biodegradable material breaks down into compost, a nutrient-rich, soil-like material coveted by gardeners and farmers. Toss scraps into kitchen collectors as you prepare meals (for convenience, BioBag makes small biodegradable, compostable bags). Create a compost system out of nothing but a section of ground, or speed things along with composting units and tools to aerate and spin the matter. Redworms, a variety of earthworms, can even do the work for you. To make a redworm composting unit (even apartment dwellers can do it), Google the term "redworm compost."

**Cookable scraps:** if your scraps are safely edible and nutritious, do what the best chefs do: make soup. Freeze vegetable, meat, and fish trimmings separately as you accumulate them; when the time is right, prepare a batch of soup or stock. (A pressure cooker is especially handy here, as is a slow cooker.) Even shrimp shells and corn cobs are packed with flavor. And don't throw out the rind from a wedge of parmesan: heat it in soup (even canned soup) for a four-star flavor boost.

**Use a wire-mesh sink strainer:** a simple wire-mesh sink strainer catches even the smallest particles of food, which can be dumped out rather than draining down the pipes. I also use a strainer for cooking tasks. Set it over a glass measuring cup, then strain hot stock, gravy, or other liquids through it, and catch seeds when reaming lemons and limes (plus you can see how much liquid you're measuring).

## Cooling Down the Hot Zone: Ovens, Microwaves, Cooktops, Grills, and Small Appliances

Cooking appliances fritter away a shocking amount of fuel. Did you know that less than 7 percent of the energy consumed by a gas oven goes to the food, and less than 14 percent for electric ovens?

Cooktops are more efficient, but even then only 40 percent of the energy consumed by a gas burner actually gets to the food. In electric cooktops, this figure averages 74 percent. But electricity is misleading: if it's not derived from a renewable source (like solar or wind), you're sucking up hidden fossil fuel, and pouring out carbon, before it even gets to your home because of the conversion process (The Footprint's Shadow on page 26 clarifies this).

Eating meals prepared at home is still more efficient than eating out, but we can do better. In this section you'll find a rundown of what to consider when outfitting your kitchen with large and small appliances. Small appliances generally consume less electricity than electric burners, and they can even beat gas burners. Because ovens are completely ridiculous when it comes to energy efficiency, *any* appliance that runs more efficiently than an oven, gas or electric, is one worth considering.

An oven's heat can warm the house on a chilly night. But most days of the year, in most places in North America, a hot kitchen leads to turning on the air conditioner or fan. To reduce the fuel you use, and to cook wisely with conservation in mind, adopt the cooking tips in Chapters 3 and 4.

## Ovens

### Gas vs. Electric Ovens

The efficiency of all ovens depends greatly on their fuel source. Electric ovens cook slightly more efficiently than do gas ovens, but natural gas burns more cleanly than electricity—unless the electricity comes from renewable

## The Footprint's Shadow: Tracking Electric Blowback

Which is greener: cooking with electricity, or natural gas? As clear as the question may seem, the answer itself is somewhat fuzzy.

What comes into your home as electricity or gas starts somewhere else, often in a very different form.

As with other fossil fuels, natural gas is considered a primary energy, meaning it's not substantially altered in order to provide consumable fuel. If you cook on a gas range, you're probably using natural gas; a small percentage of people, usually in rural areas or where natural gas lines aren't available, may opt for liquid propane. These carbon-based fuels, as one would expect, release more carbon dioxide into the atmosphere than noncarbon fuels. But that's where the green distinction between gas and electricity gets tricky.

Electricity itself is always created out of some other energy source, so it's considered secondary energy. The carbon-based energy used to create electricity is what I call the "footprint's shadow"—you don't really see it, but it's there. When electricity is generated from coal plants, for instance, it averages about three kilowatts of fossil fuel to generate one kilowatt of electricity. That's a big shadow, bigger than

*(continues)*

resources, like solar or wind. If your electricity is low-impact or renewably sourced, electric ovens make more sense than gas ovens. But even with their lower efficiency, gas ovens are still better picks if your electricity is gener-ated as secondary energy (like from coal plants). Natural gas isn't an option in all locales, so you may have to opt for electric.

## Wall Ovens

Ovens perform the same whether they're part of a range or separate from the cooktop. But, do you really need a *double* wall oven? Sure, designer kitchens look sexy with a wall of ovens, but realistically, how often do you cook with both ovens simultaneously? A toaster or countertop oven saves fuel, comes with a smaller manufacturing footprint, doesn't heat up the house, and still gives you an additional oven and broiler. (More on this on the next page.)

(The Footprint's Shadow: Tracking Electric Blowback, *continued*)

straight natural gas alone. Electricity's shadow grows or shrinks in size according to the forms of primary energy used, like nuclear fission, hydroelectric dams, petroleum oil, and sustainable sources. Many people, I suspect, have no idea what type of fuel makes the electricity that fuels their homes.

But there's good news. Commercial electricity is increasingly being sourced from wind, solar, and geothermal suppliers. So even if you're plugged into the grid, you may be able to choose the primary source of your electricity and avoid fossil fuels altogether.

Comparing electricity to gas is much like comparing apples to oranges. One kilowatt of electricity may have a three-kilowatt shadow, or no shadow at all (if you independently generate your own solar, wind, or geothermal power). Electricity from your power company may be a greener choice than natural gas if it's sourced from wind or sun. Coal-generated electricity emits more carbon than does natural gas, but both spew $CO_2$.

Which is greener: cooking on an electric stove, or a gas stove? The best solution is to reduce emissions in every way we can—from the fuels we choose, to the amount of energy we consume.

## Gas Ignition Options

Like the Olympic flame, a gas pilot light burns continuously; but the range is just hardware, not a symbol of the world's fastest men and women. Electronic ignitions come standard with most new gas ranges, meaning no fuel to waste. If an electrical power outage occurs, fire up the burners with matches.

## Convection Option

Use a convection oven and you'll save time and fuel by about 25 percent. But not all convection is true convection. Check the section on page 53 for how to cook with convection.

## Self-Cleaning Ovens

Even if you never use the self-cleaning function, it's a worthwhile feature. Self-cleaning ovens are better insulated than standard ovens, so less heat pours into your kitchen. If you do use the self-cleaning feature, start it when your oven's already hot, like after roasting, and limit frequency to no more than once a month.

*Oven Cleaning Tip:* Oven liners, heat-proof mats that rest on the floor of your oven, can help conserve fuel. Instead of running the incendiary self-cleaning feature, you may need to just clean the liner, which captures spills (but not splatters on the sides of the oven). But don't cover the racks with a mat or foil: you want the heat to circulate, and covering the racks prevents this.

## Toaster and Countertop Ovens

If you do nothing else to green your cookprint, switch from a conventional oven to a toaster or countertop oven. Anytime you can bake, roast, or broil in a smaller oven, you're cooking greener. Plus, compared to conventional ovens, small ovens preheat and cool down in almost no time at all, and they don't heat up the entire kitchen. Cheaper models may cost less, but if they don't perform well, you won't be able to bake, roast, or broil in them in the same way you use a conventional oven. Look for ones with good consumer reviews based on cooking accuracy, size, number of racks, and convection and broiling features. The unit's actual footprint, the space it takes on your countertop, can vary. My toaster oven's compact footprint measures just 12 x 15 inches, yet even the 0.5 cubic foot interior is big enough to roast a chicken, bake casseroles and cornbread, char chiles, and broil garlic bread. Some models have contoured walls to accommodate a 12-inch pizza, and other bells and whistles appear with every new model. *Consumer Reports* tests these ovens periodically, so look to them and other cooking sources for the most current reviews.

### Microwave Ovens

Cooking with microwaves can reduce energy consumption by as much as two-thirds. As cooking results go, microwave ovens have their pros and cons, so flip to page 57 to discover green details and how to microwave with tastier results.

### Hybrid Speed Ovens

These "speed ranges" combine microwaves for fast cooking with radiant heat or halogen for browning. Some cooks have adapted to them, so this technology may work for you. But if you don't think you'll use the speed features, they're not worth the extra expense or manufacturing carbon.

### Solar Ovens

Although solar ovens are obviously not standard kitchen appliances, they are energy free, and you can use them for everything from cooking rice to baking desserts (see page 42).

## Cooktops

### Gas Burners

As with gas ovens, avoid pilots that burn continuously. Gas burners are less fuel-efficient than electric burners because of the amount of heat lost between the flame and the pot. Sealed gas burners are easier to clean and about 5 percent more efficient than conventional burners. Gas units need a ventilation system to vent fumes to the outdoors; updraft hoods are more efficient in this respect than downdraft ones.

### Electric Burners

Glass cooktops with radiant or halogen elements are better options than exposed coil elements. Avoid solid disk burners.

### Induction Burners

Induction burners make me jump for joy. Electromagnetic energy drives these burners, though you need to use pans made of ferrous materials (like

# Portable Induction Burners: Energy-Efficiency in a Box

Full induction cooktops, with fully loaded price tags, are creeping into the marketplace. But a single induction burner can supplement your existing gas or electric cooktop more affordably.

Induction cooking works by sending a magnetic field through ferrous metal (as in cookware made of iron, steel, or a combination). The reaction creates heat, and it's this heat that cooks the food. The heat is created from within the pan's own material; think friction and fast-moving, excited molecules (like the heat generated between your hands when you rub your palms together).

The result: a near-instant transfer of energy, with efficient absorption of over 90 percent of this energy (compared to around 40 percent efficiency with gas burners, and 74 percent with conventional electric burners). Plus, the cooker's surface stays cool, very little heat is released into the kitchen, and the food can actually cook more quickly. Because the cooker surface stays cool, absorbing heat only from the cooking vessel, it's easy to clean (no stuck muck). Plus, with a nifty portable unit, I can cook anywhere there's a plug, even outdoors.

The first time I boiled pasta or fried steaks on the induction element, I noticed the differences from conventional electric or gas cooking right off the bat. The water boiled sooner, and the fry pan reached perfect searing heat in a flash. Plus, I had instant control; when I turned the dial from high to low, the unit powered down to the lower setting immediately. No waiting for a hot gas or electric element to slug down in speed. And you can maintain constant simmering and very low temperatures better with induction.

With induction, there's no learning curve (unlike with microwave ovens or speed ovens). You do need to check your cookware: only ferrous metals are induction-compatible, but fortunately this includes everyday iron-and steel-based cookware. Basic rule: if a magnet sticks to the pan, it will work with induction. (This eliminates glass, copper, and purely aluminum pans.) Most portable units run between 1,400 and 2,000 watts, on a standard 120-volt power outlet.

cast iron or stainless steel). They consume less than half the energy of standard coil burners, and they're superior in cooking performance to gas or electric burners. Single induction burners make handy portable appliances, and some conventional cooktops add a separate induction burner.

## Small Appliances: The Good, the Bad, the Worthless

*Consider this:* If we all had to pay yearly taxes and registration fees on our appliances, like we do on vehicles, how many beaters, mixers, cookers, and other electric gadgets would you keep in your appliance garage?

Thrift shops teem with used appliances, ones that do everything from shoot your salad to shake your dressing. Such small appliances once lived bright and shiny lives in the kitchenware aisle of Target or Wal-Mart. They go home with Aunt Bee as an easy Christmas gift for niece Tiffany, or are spontaneously shoved into carts based on their promise of making busy lives easier and better. But when the romance dies, or the products fail to deliver, off they go to the Thrift Shop's Appliance Graveyard, a slightly better fate than the landfill but not by much. Eventually, they'll probably wind up in a landfill anyway, even if someone rescues them for one last chance at kitchen bliss.

While some small appliances are cookprint nightmares, others offer true green value. The key is using them often enough to offset their manufacturing footprint, and selecting ones that will, when compared to conventional methods, cook with less fuel, less water, or both.

The best small appliances waste very little fuel. Most function as closed systems, with tight lids and heating elements that channel almost every bit of energy into the food, so the room stays cooler, the food cooks efficiently, and natural moisture (with nutrients) is retained. But that reasoning alone doesn't justify one-trick-ponies like muffin makers, electric fondue pots, and flowing chocolate fountains.

On the other hand, you don't have to be a caveman in the kitchen. Some plug-ins offer real benefits, and this section zeros in on the greenest ones to consider.

## *Weighing In on Wattage*

How much electricity does a small appliance use? Check the wattage on the box (or on the back, bottom, or nameplate). Wattage is the *maximum* power drawn by the appliance, so changing settings (like running on low rather than high) can reduce the amount of energy pulled.

Measure with a meter: pick up an inexpensive Kill-a-Watt electric usage meter. Plug your unit into the meter, and the meter into the wall socket. It registers the amount of electricity pulled, both when actual usage occurs and during standby or vampire modes. Some models even calculate your actual usage in dollars (enter in your rate and it does the math). Remember: unplug appliances when not in use to avoid phantom loads.

With appliances that cycle on and off, like refrigerators and ovens, you need to guesstimate how frequently the "on" cycles into action. The U.S. De-

## Kitchen Fans: What to Look For

### *Exhaust Fans*

A kitchen exhaust fan sucks power, but it also sucks up fumes and fine airborne grease particles that gum up surfaces. Fans that vent to the outside are more efficient than nonventing models, which recirculate air through charcoal filters (like the downdraft systems of island cooktops). Be sure to consider the CFM—cubic feet per minute, the amount of air the system can pull through. A low CFM (150 to 200) won't make much difference, and designer hoods with CFMs of 500 to 1,000 or more may be overkill. I use an over-the-range microwave oven with a built-in exhaust fan of 300 CFM. It's perfectly fine for home cooking without being a fuel-hog.

### *Ceiling Fans*

Ceiling fans are a low-energy alternative to air conditioning for cooling down a kitchen. Because the motors give off some heat, run them only when you're in the room to benefit from the air circulation. When a gas burner's on, leave a ceiling fan off if it disrupts the heat flow from source to pan.

partment of Energy's estimate for refrigerators is about a third of their maximum wattage measured over time (see the sidebar on page 57). For greater accuracy, plug your refrigerator into an electric usage meter and check the measures over a one-week period in summer and a week in winter, then average them out.

## Favorite Small Appliances

The toaster oven, portable induction burner, and electric tea kettle (see the sidebar on page 57) top my list of good green appliances. Other picks are based on the foods I cook most, and your list may be different. What makes a small appliance *green* is its ability to cut back on everyday fuel consumption, enough to justify the unit's manufacturing cookprint. Follow this section's power guidelines and check the wattage when making your choice.

### Rice Cookers

Decades ago my Korean mother switched from cooking rice in a pot to using an electric rice cooker, and our family's never looked back. Energy-wise, it's the most efficient way to cook all kinds of rice, using two to three times less fuel than conventional methods, and it doesn't heat up the kitchen. It shuts off automatically, keeps rice steaming hot for long periods, and reheats even cold rice to perfection. Some models let you steam vegetables and other grains, but if you don't enjoy rice on a regular basis, this may not be the appliance for you. (Pressure cookers cook rice with about 50 percent less energy than a cooktop.) *Tip:* regardless of the method used, soaking rice for 15–30 minutes before cooking shaves off 7 to 18 percent of cooking fuel.

### Slow Cookers

Slow cookers are getting makeovers. They've always been more energy-efficient than the oven or the cooktop, and now some models offer removable inserts (the crock) that can withstand stovetop cooking: no need to dirty another pot or pan just to brown chicken or roasts. Most crocks can also go into microwave ovens, refrigerators, and dishwashers. Slow cookers prove even more valuable in summer than winter, since they don't heat up the kitchen. They cook more than just stews and soupy things, like Cuban black bean salad, chili con queso, spicy roasted pecans, garden tomato pasta sauce, and dreamy lemon cheesecake (find other examples in this book's recipe section). Slow cookers consume less electricity than an incandescent light bulb. And for cooking in off-peak hours (like overnight), this appliance's long, slow cooking method is ideal.

## Formula for Estimating Energy Consumption

(Adapted from the U.S. Department of Energy)

Use this formula to estimate an appliance's energy use:

$$\text{Wattage} \times \frac{\text{Hours Used Per Day}}{} \div 1000 = \frac{\text{Daily Kilowatt-hour (kWh)}}{\text{consumption}}$$

(1 kilowatt [kW] = 1,000 Watts)

Multiply this by the number of days you use the appliance during the year for the annual consumption. Calculate the annual cost to run an appliance by multiplying the kWh per year by your local utility's rate per kWh consumed.

**Note:** To estimate the number of hours that a refrigerator actually operates at its maximum wattage, divide the total time the refrigerator is plugged in by three. Refrigerators, although turned "on" all the time, actually cycle on and off as needed to maintain interior temperatures.

### *Examples:*

Window fan:

200 Watts × 4 hours/day × 120 days/year ÷ 1000= 96 kWh × 8.5 cents/kWh= $8.16/year

Personal computer and monitor:

(120 + 150 Watts × 4 hours/day × 365 days/year) ÷ 1000= 394 kWh

394 kWh × 8.5 cents/kWh= $33.51/year

**Wattage:** You can usually find the wattage of most appliances stamped on the bottom or back of the appliance, or on its nameplate. The wattage listed is the maximum power drawn by the appliance. Since many appliances have a range of settings (for example, the volume on a radio), the actual amount of power consumed depends on the setting used at any one time.

If the wattage is not listed, you can estimate it by finding the current draw (in amperes) and multiplying that by the voltage used by

*(continues)*

(Formula for Estimating Energy Consumption, *continued*)
the appliance. Most appliances in the United States use 120 volts. Larger appliances, like clothes dryers and electric cooktops, use 240 volts. The amperes might be stamped on the unit in place of the wattage. If not, find a clamp-on ammeter—an electrician's tool that clamps around one of the two wires on the appliance. Take a reading while the device is running; this is the actual amount of current being used at that instant.

When measuring the current drawn by a *motor*, note that the meter will show about three times more current in the first second that the motor starts than when it is running smoothly.

Many appliances continue to draw a small amount of power when they are switched off. These "phantom loads" occur in most appliances, such as VCRs, televisions, stereos, computers, and kitchen appliances, and will increase energy consumption a few watt-hours. These loads can be avoided by unplugging the appliance or using a power strip and switching off the power strip to cut all power to the appliance.

### Typical Wattages of Various Appliances

Here are examples of nameplate wattages for various household appliances:

Clock radio = 10
Coffee maker = 900–1,200
Clothes washer = 350–500
Clothes dryer = 1,800–5,000
Dishwasher = 1,200–2,400 (the drying feature greatly increases energy consumption)
Fan, ceiling = 65–175
Fan, window = 55–250
Fan, whole house = 240–750
Microwave oven = 750–1,100
Refrigerator *(frost-free, 16 cubic feet)* = 725
Television (color, 36 inches) = 133
Toaster = 800–1,400
Toaster oven = 1,225
Vacuum cleaner = 1,000–1,440
Water heater *(40 gallons)* = 4,500–5,500

### Immersion Blenders

If I were trapped on a desert island and could have only one chopping appliance, I'd choose the immersion blender. I call it "my iPod of the kitchen." Chopping by hand will always be the most fuel-efficient method, but recipes routinely call for blenders and food processors, and an immersion blender plugs cooks into a greener option.

How does the immersion blender (or "hand blender") compare with a regular blender, mixer, and food processor? It's compact and portable, and well suited to smaller, everyday-size quantities. With the chopping and whisking attachments, a handblender can chop, pulverize, granulate, mince, puree, and whip—everything from soup to nuts, or salsa to stir fry, and whipped cream to coulis.

The best immersion blenders perform powerfully, yet draw far less energy than full-size blenders or food processors. Rinse out the hand blender in seconds, and stick it in a drawer when not in use. You can't say that about stand mixers, powerful hunks of metal that weigh more than most toddlers (about 25 pounds), with nearly enough horsepower to steer a boat. (On the other hand, a good-quality stand mixer will last a lifetime or more with proper maintenance.)

If a rice cooker, breadmaker, or any appliance won't get a lot of use in your home, or reduce fuel compared to conventional cooking methods, then don't get seduced by sales pitches. Specialized appliances are only bargains if they shrink your overall cookprint.

## Lighting Up the Dry Zone

Countertops as well as floors constitute the most active dry zones in a kitchen, and cabinets, drawers, and storage areas are also considered dry areas. Before remodeling your kitchen, dig deep to find the latest, and truest, green materials. Like food, kitchen materials that are local and sustainable beat out imported and nonrenewable, but you may have to settle for trade-offs. (Bamboo, as mentioned, has sustainable appeal, but most is shipped from China.)

You'll also make better use of your workspace if the area is well lit, and converting to energy-efficient lighting is far less demanding than replacing countertops and cabinets. Kitchens need two main types of lighting: overhead and task lighting. With proper under-cabinet lighting, you may actually prefer to work without flicking on the overhead lights. The fewer the lights, the less energy consumed, and the cooler the kitchen's temperature will be.

## Gold-Star Green Tool: The Electric Teapot

Mention an electric teapot, and most Americans are clueless. But to the tea-loving British, electric kettles are everyday appliances. The electric kettle can be more energy-efficient than either a gas or electric burner, and here's why:

The best electric teapots heat from within, not by sitting on a hotplate. Most boil 6–8 cups of water in about 5 minutes, several minutes quicker than on a burner. You'll be greener if you heat only the amount you plan to use. A heating element extends upward, inside the pot, so it's surrounded by water. The element becomes very hot, very quickly, and directly transfers that energy to the water around it. Nearly 100 percent of the energy goes to heating the water, almost no heat escapes into the air, and the speed reduces energy lost during heating, unlike a pot of water on the cooktop. Besides the stay-cool handle, the kettle exterior stays cooler, and the lid prevents energy and steam from escaping. The kettle shuts off automatically, and the heating element cools down far more rapidly than the gas or electric burners of a stove.

Most kettles hold 6–8 cups, not enough to replace a 4-quart pot of pasta water. But oh, the other things you can do with one! Chapter 4 shows how to passively blanch vegetables and cook creatively with boiled water from an electric kettle. Besides unconventional uses, electric kettles are *absolutely* more efficient for beverages, instant soup, rehydrating dried mushrooms or tomatoes, or when you need boiling water to pour over couscous or bulgur wheat, for instance.

An electric kettle may not be essential, but it does what it's designed to do remarkably well. Still, if a wired kettle will be just another dusty, tired appliance in your home, then don't bother buying one—or at least give it to your nearest tea-sipping, tree-hugging, chipper British neighbor.

Speaking of heat, halogen lights may be trendy and efficient, but they're notoriously hot. For under-cabinet lights, I prefer cool-burning, energy-efficient fluorescent tubes. Narrow, short lengths can be spaced every couple feet, and they come in full spectrum, without the harshness of old fluorescent lighting. CFLs, compact fluorescent lights, are super-efficient, but they contain trace amounts of mercury. Check with your local recycling program or big-box hardware store about proper disposal.

The greenest option, LEDs (light emitting diodes), are coming down in price, and ramping up in models. LEDs work by clustering a series of small bulbs together (single bulbs are typically used in pen lights or instrument panels). They're ultra-efficient, cool burning, and last ten times longer than compact fluorescents (without mercury). They don't radiate a broad swath of light, but they're terrific for task lighting (over cutting boards and sinks) and desk lamps, and are commonly used for street lamps, traffic lights, garden lights, and car headlamps.

## The Outdoor Zone: Greener Grilling and Emission-Free Solar

*Nationwide, the estimated 60 million barbecues held on the Fourth of July alone consume enough energy—in the form of charcoal, lighter fluid, gas, and electricity—to power 20,000 households for a year. That one day of fun, food, and celebration, says Tristram West, a research scientist with the U.S. Department of Energy, burns the equivalent of 2,300 acres of forest and releases 225,000 metric tons of carbon dioxide.*

—Sierra *magazine*

According to a 2005 survey, 81 percent of U.S. households own at least one grill (most households own multiple grills), and more than 15 million new grills were shipped in 2005 alone. That's a hefty hunk of burning metal.

By 2007, half of all grill owners in America owned a charcoal grill, up from 42 percent in 2003. Overall, gas grills still dominate at 67 percent, but the rise in charcoal grilling comes at a bad time environmentally. (Electric models account for only 7 percent of owned grills.)

Though grilling with natural gas, propane, and electricity all consume natural resources and release airborne toxins, none of these methods do as much

## Share the Wealth: Throw an Appliance Swap

In our household, certain appliances rarely get a rest. Like the rice cooker, toaster oven, and hand blender. But after the first month, I rarely used the quesadilla maker, and even my panini grill gets dustier by the day.

If you have appliances shuttered away, or ones you're keeping on hand for a rainy day, dust them off. Call up your friends, neighbors or relatives. Do a swap. Share the wealth; let them enjoy the novelty or convenience for a while. Your lifestyle may no longer include an electric margarita maker, but your cousin might have just the right use for it. I recently passed off my KitchenAid stand mixer to my teenage niece, who's just discovering the joys of baking. It's a permanent loan: it's all hers unless one day I need to borrow it back.

Charitable resale shops are another way to repurpose small appliances for a greater good. Keep the original box and manual to make them more desirable and perhaps fetch a better price.

**First Option for Shoppin': Reuse for a Smaller Cookprint.** Adopting orphaned appliances prevents bringing yet another unwanted toaster or slow cooker into this world. Before buying a shiny new appliance, consider rescuing a pre-owned unit from a thrift shop or garage sale.

damage as charcoal. The mixture of summer heat and charcoal emissions creates a mix of ground-level ozone that's intensely toxic in big, hot cities on July Fourth and other peak-grilling days.

Admittedly, I'd have a hard time living without the smoky flavor of grilled foods. Plus, grilling offers certain tangible benefits: fewer pots and pans to wash, and keeping the house cooler than indoor cooking, for instance. But even though I authored *The Stubb's Bar-B-Q Cookbook,* I'm rethinking how I grill, and doing it completely without charcoal.

Die-hard grillers can make a green difference by following these tips:

## Choose a Chimney over Lighter Fluid

When the VOCs (volutile organic compounds) in lighter fluid evaporate and meet sunlight, they party with other air pollutants and leave a toxic mess known as ground-level ozone. Unlike the atmospheric ozone that protects us, this low-lying ozone layer literally chokes us up, especially anyone with heart and lung conditions, children, and folks working or exercising outdoors. Plus, lighter fluid leaves toxic residues on food. Avoid instant-light charcoal briquettes too, which contain lighter fluid. Instead, fire up a charcoal chimney. The device looks like a coffee can with a handle, and it uses a wad of newspaper to ignite the charcoal inside without petroleum products. An electric charcoal starter also works better than lighter fluid; place the coil under the charcoal, and remove it once the charcoal is lit.

## Limit or Lump Your Charcoal

We love the flavor it adds, but charcoal is, not surprisingly, on the black end of the green scale. Made from wood (rarely sustainable, and usually adding to deforestation), commercial charcoal is manufactured by burning wood in hot kilns to 500+ degrees F. So even before it's bagged and shipped, it's already spewed $CO_2$ gases and solid residue into the air. On the backyard grill, charcoal continues to release carbon monoxide, soot, and other particles that, under the summer sun, build ground-level ozone into an ugly, unhealthful layer. But if you've got an insatiable charcoal Jones, switch to lump charcoal (and a charcoal chimney). Lump charcoal comes from tree limbs, sawmill ends, or dried lumber scraps and is preferred over compressed and extruded briquettes, which typically include wood waste, additives, binders, and coal dust, making their emissions an especially unsavory form of air pollution.

## Grill with Natural Gas, Propane, or Electricity

Natural gas, propane, and electricity avoid the soot, carbon monoxide, and heavy-duty particulate matter that charcoal creates. Plus, you can grill indoors or on patios with certain portable electric grills; most let you plop the grill pan and grates right into the dishwasher, and they're really fun for tabletop grilling of skewered foods, mixed vegetables, or Asian-style dining.

## Clean Your Grill with Tools

You don't want a dirty grill contaminating your food, and burning is an effective way to kill any living thing. But if you scrape off most of the residue with a sturdy wire brush before firing up the grill, you'll burn less fuel to get it clean and send less smoke into the air. A long-handled tool known as a Grill Floss (unfortunate name but it works) can scrape each rod of the grate from end to end, even in hard-to-reach places like under the rod and in corners. For an occasional scrubbing, SoyClean makes a spray on/wipe off cleaner, and many natural cleaning products work well and rinse off with water.

## Keep the Door Closed

Put less demand on the AC by limiting trips to and from the kitchen. Organize everything and take it out all at once. Trays and rolling carts help. An ice chest with reusable chill packs keeps food and drinks chilled while you're grilling.

## Renovate and Repair

More than 15 million grills ship annually. Eighty percent of all grills are priced under $300—cheap enough to punch our consumer buttons. Most of these grills aren't very complicated (they're basically metal boxes with plumbing). Periodic maintenance may be all that's needed to restore the looks and performance of your old grill. Before rushing out to buy a new model, scrub up your old one and replace parts, like igniters or grates, or give it a new paint job. New grills may entice with their sleek shiny looks, but ultimately they may not work better (or even as well) as the grill you're already using. Restoring a grill eliminates more manufacturing pollution, materials, and fuels to transport a new grill from overseas (or wherever it's made).

## Buy Locally

In Texas and most other states where grilling is a religion, small companies weld their own lines of consumer grills, including propane-capable ones. Most mass-produced grills are made in China and other foreign nations. If you are in the market for a new grill, see who's putting their local or domestic stamp on the grills in your area.

## Try a Solar Cooker

Because solar cooking is totally emission-free, you can't get much greener. Admittedly, solar cookers don't create the same effect as grills, but if being outdoors and cooking green are dual objectives, check into the various solar cookers available. You can roast a brisket; bake casseroles, cornbread, and cookies; and cook just about anything you'd put in a slow cooker (they have similar temperature ranges). One model has gone tech: the Tulsi Hybrid Solar Cooker includes a low-wattage electrical back-up system, a handy option when that unexpected cloud rains on your parade, or you just need to speed up the process. Books and online resources offer plans for making solar ovens, buying ones, and specific recipes and cooking tips.

# Cleaning Your Zones

One of the best tips I ever read about cleaning is to *avoid* it: Reduce surface areas for dust and grime to cling to, especially in the kitchen, and you'll have a more clean-free environment.

- Open cabinets look cool, but they let airborne grease collect inside cabinet surfaces and on their contents. Instead, opt for cabinet doors with glass panels.

## Use Cloth Napkins Every Day

Before paper napkins were invented, people used—and reused—cloth napkins. Napkin rings kept the used napkins tidy, and in some cases, identified whose napkin belonged to whom. Organic cotton or other renewable cloth napkins also add color and a degree of civility to dining. To clean, simply toss them into the weekly laundry (napkin rings optional).

- Flat surfaces are easier to clean than sculpted ones, which is why smooth cooktops and sealed burners are preferred. Manually wipe up spills when they're fresh, to prevent cooked-on gunk and reduce total cleaning.

- Kick the antibacterial soap habit. Don't use soaps and household cleansers with "antibacterial" properties. We're creating super-germs: each new generation of bacteria grows more resistant to the overabundance of antibacterial chemicals. As this happens, antibacterial claims grow unsupportable; you may be operating under a false sense of security. Plus, antibacterial chemicals make no distinction between good and bad bacteria, and the good bacteria have a necessary place in our ecosystem. Regular soap works best.

## Recycling Large Kitchen Appliances

The best way to recycle or dispose of refrigerators, stoves, and ovens depends on where you live. Check with local government and area recycling programs. In general:

- If your appliance works, consider donating it to Habitat for Humanity, Goodwill, or another charity for resale, but check first to see what they accept. Some charities pick up nonworking appliances if they can repair them.
- If your clunker appliance is an energy hog, let it rest in piece, or pieces as the case may be. Older appliances (ten years or more) typically contain ozone-depleting and toxic materials, making them ill suited to landfills. Your city or area disposal service can advise on the best options in accordance with federal guidelines.
- When buying a new appliance, ask the store whether it will recycle your old unit or dispose of it properly.

Some utility companies sponsor bounty or incentive programs, including rebates for Energy Star appliances, and will also pick up and properly dispose of your old appliance.

- Products labeled "safe for septic tanks" are on the good list. Avoid products that have antibacterial, chlorine, phosphate, dye, or artificial fragrance on their labels.

- Use green cleansers, either store-bought or home made. Linda Mason Hunter and other authors have written books on green-cleaning everything from cars to cats. White vinegar attacks grease and dirt and is safe for people and pets. Try this all-purpose formula: 1 part white vinegar, 1 part water. Spray or rub on glass, tile, mirrors, countertops, ovens, refrigerators and freezers (interior and exterior), stoves, and floors, and wipe clean. For tougher grease, add a squirt of biodegradable dish soap, which acts as a surfactant.

- Choose cellulose (made from wood pulp) sponges over petroleum-based polyurethane. (Sea sponges would be another option, if they weren't being depleted in nature.)

- Use cloth dishtowels instead of paper towels most of the time. When you do need a paper towel, opt for ones made with recycled post-consumer waste, and processed chlorine-free (PCF).

- Place a mat or skid-free rug on areas prone to spills, like by the stove, sink, and the countertops where most chopping happens. Shake out or vacuum the rug, and wash periodically, instead of washing the floor as often.

# THREE

# Ovens: Improving Your Mileage

If you make *no* other changes to the ways you cook, stop using your oven.

Ovens are the Humvees of the kitchen. According to the Department of Energy, about 87 percent of an electric oven's heat is wasted, absorbed not by the food but by the oven walls or dissipated into the room. Gas ovens lose a staggering 94 percent of the fuel, through the walls, flue, and inefficient combustion. Contrast this with stovetop cooking, which makes efficient use of 70 to 80 percent of the fuel, depending on whether it's a gas or electric burner. Yet, most of us rarely think twice when turning on the oven.

Does this mean we should stop baking breads, cookies, and cakes? Or that we should say ta-ta to Thanksgiving turkeys, roasted butternut squash, or broiled salmon? Not necessarily. Put lasagna, baked potatoes, and baked, roasted, and broiled dishes on an occasional, cold-weather menu. We can also tweak our cooking to make better use of an oven's fuel, and opt for fuel-efficient alternatives, like cooktop, toaster oven, slow-cooker, and microwave oven methods.

## How Ovens Work

Think of the way a home's central heating system works. Say it's a chilly December day, and you've been away with the heater off. As soon as you walk in the door, you set the thermostat to a cozy 70 degrees F. The unit rumbles into action, burning fuel continuously until the temperature registers 70 degrees. After that, the system cycles on and off, kicking in only when the room air drops a few degrees, and cycling off again when it hits the 70-degree mark.

Ovens work the same way: they burn fuel continuously until they reach their mark. Whenever the temperature drops below that setting, the heat kicks in again.

## Weather-Sensitive Cooking

*Fall Canceled After 3 Billion Seasons—WASHINGTON, D.C.*
*Fall, the long-running series of shorter days and cooler nights,*
*was canceled earlier this week after nearly 3 billion seasons*
*on Earth, sources reported Tuesday.*
                    —The Onion, *November 7, 2007*

With global weirding, climactic norms no longer exist. Where I live, temperatures can swing 30 degrees in a matter of hours, sending us scurrying from al fresco lunching in T-shirts, to supping on stews in sweaters at night. I once planned dinners with a seasonal nod, but now I'm juggling menus not only day by day, but meal by meal. Lasagna in December just doesn't seem appetizing when it's 90 degrees outside.

Outdoor temperatures affect more than just appetites. Cooking methods directly affect the home heating/cooling environment. Oven roasting or baking while the air conditioner runs is obviously energy foolish, yet many of us can plead guilty. On the other hand, an oven that warms the room and offsets the central heater on a chilly day makes some sense: you're stretching the fuel you use, while gaining practical and gustatory fulfillment at the same time (even then, ovens are drastically inefficient). My advice: be flexible. Select a cooking method to match the current weather, not just the season. And unless you have a really good reason to fire up your oven, just don't do it—especially on hot weather days.

If your home is well insulated, with thermal windows and few gaps where cold air can seep in and hot air escape, then the central heater doesn't need to cycle on and off very frequently. Likewise, a well-insulated oven, with no leaks and a good seal, retains heat better, and uses less fuel, than a poorly insulated one. (That's why self-cleaning ovens are more energy-efficient: they fire up to 850 or even 1,000 degrees, so they're equipped with better insulation and stronger seals than non–self-cleaning ovens.)

When you cook in an oven (or broiler), you're burning fuel in three ways. A batch of cookies may bake in only "10–12 minutes," but you're expending some 10 to 20 minutes of fuel just to preheat the oven. The process of getting a cold oven up to temperature consumes a big chunk of cooking fuel. Maintaining the cooking temperature is less fuel-consuming, because the oven will cycle on and off. However, once the oven's turned off, all that powerful heat collected inside it is wasted, left without a purpose. Plus, throughout the entire process, from the moment you turn on the oven until the time when it's completely cool again, ambient heat radiates from the oven into the room, one of the inefficient by-products that comes with just making an oven hot.

## What to Do with Your Oven

Treat your oven as a luxury. Turn it on only in cold weather, and when you do use your oven, maximize the fuel efficiency. Get into the habit of cooking with all three stages of an oven's fuel: preheating, active cooking, and cooling (or post shut-down, up to the point when there's no longer any appreciable heat). In this chapter, you'll discover techniques you can apply to all types of everyday fare.

Commit to oven multitasking, too: make a pact to always cook two or three dishes at the same time, even if it's just roasting garlic or toasting nuts. (A single oven cooking three dishes simultaneously can be as efficient or more than firing up three burners.) In almost all cases, shutting off the heat several minutes early works fine, especially with thick, heat-retaining cookware.

Will all of these strategies work with your oven? Yes, but you may need to tweak them a bit, since every oven is different in size, shape, and materials; and as we'll dive into later, so are the pots and pans we cook with. But overall, the strategies are sound, flexible, and easy to integrate into one's daily routine.

## Blue Oven Strategies = Green Results

This passive cooking technique may go by other names, but for me "blue ovens" sums it up best. On its most basic level, you simply turn off the fuel early and let the accumulated heat finish cooking the food. You can do this with most casseroles, without changing cooking temperature.

A different blue oven strategy cooks at a higher than normal temperature, but only for a short time. Then the fuel is shut off and the door left closed. It's good for baking or roasting large, solid foods (a roast, thick fish steaks, or whole eggplant, for instance). The initial blast of heat does two things: it kills surface bacteria, and it browns the exterior, creating tasty caramelized flavors.

With solid hunks of protein (like roasts), food-borne bacteria tend to reside on the exterior, so a fast hot blast eradicates the nasties, leaving the interior to cook gently and gradually, using just the oven's residual heat. For example, Blue Oven Roast Beef draws on just fifteen minutes of active fuel plus preheating fuel (conventional methods burn up to two hours of fuel). If you cook another dish at the same time, like potatoes, your cookprint grows even greener.

To test-drive blue oven cooking with your own kitchen Humvee, flip to these recipes:

- Blue Oven Rare Roast Beef, page 196

- Short-Cut, Passive Lasagna, page 203

- Spatchcocked Roast Chicken, page 181

## More Blue-Oven and Companion Strategies

To consume less fuel, and to *waste* less fuel when you cook, adopt these other oven strategies.

## Power Down Early

Roast or bake as you normally would, but cut off the oven's heat five to fifteen minutes early, then let the dishes or baked goods passively cook in the hot oven until done (the oven's temperature won't change much during that time). Baked potatoes can finish in a powered-down oven, and even better, they can also start in a cold oven. Experiment with your own recipes to see which ones work best. What's the worse that could happen? The final product might be a little undercooked, which you can fix with added cooking in the oven or microwave; or it may not be quite as brown, but at least you'll know if power-down cooking can be applied to that particular dish, and you'll save fuel every time you make it. Try powering down 10–15 minutes early for longer-cooking casseroles, roasts, and baked potatoes, and five minutes or so for shorter-cooking dishes.

## Cold-Start Baking: Ditch the Preheat

Almost every recipe begins with "Preheat the oven to. . . ." But this golden rule is one to break, or at least give a new green sheen to. You don't always have to preheat an oven. Many recipes work just fine started in a cold oven. Preheating is done to ensure consistency, and for some dishes it can be critical to success (mostly in baking). But a flexible cook can adapt many recipes with great success. I've had good results with casseroles, lasagna, meatloaf, winter squash, and baked potatoes, to name a few. But for pastries and yeast doughs, you're better off following the recipe as written, and going greener by cooking other dishes in the oven at the same time (more on this later). Of course, there's an exception to every rule: Fleischmann's Yeast has developed recipes that start in a cold oven. The Cold-Oven Clove and Crystallized Ginger Cake on page 244 also bakes without preheating.

## Preheat When Ready

Preheat when *you're* ready, not just because the recipe says so in the first step. Often it takes longer to prep a dish than it does for the oven to preheat. If you turn on the oven too soon, you're wasting fuel. Be mindful of how long it takes your oven to preheat, and how long it may realistically take you to prep the dish. Certain recipes really must fly directly into a hot oven, but most can wait a few minutes if the oven's not quite up to speed.

## *Multitask: Practice Simultaneous Baking*

Cooking more than one item in the same oven stretches fuel. To make sure everything cooks evenly:

- Stagger foods to let air circulate. Put them on separate racks, not directly above each other.

- Shift pans from upper to lower rack and back to front, rotating them around once or twice. But don't overdo it: opening the door loses heat, which raises fuel usage.

- Use the convection setting, if you have one, to keep air moving.

## Sniffing Out Simultaneous Baking

Can—or should—you bake a savory dish at the same time as a sweet treat? Experts are split on the issue. Bakeries and restaurants do it all the time, but some professionals avoid it, especially when perfection is paramount.

Home cooks, though, can get by just fine without separate ovens: cooking sweet and savory dishes simultaneously can be done, and it's really a judgment call. Curried lamb and chocolate cake are probably not good oven companions, but rosemary-chicken and apple cobbler should work just fine (rosemary and apples marry well in their own right). You shouldn't suffer complaints if whole-wheat bread and carrot cake commingle in the same oven, either. I've roasted potatoes with pumpkin pies, and no one could tell they were oven-mates. But the aromas from a garlicky lasagna might be just too discernable in a flaky almond tart.

My advice: use common sense, and take the plunge to test things out—especially if dishes are to be part of the same meal. What's the worst that can happen? Most people aren't sensitive to minor hints of flavor. If you don't tell anyone that two dishes were baked at the same time (with aromas wafting through the house), they may never know. If they do detect a disharmonious mix of flavors, then what the heck: it's just one meal in a lifetime of eating. If you don't experiment, you'll never know.

- Choose dishes that cook at, or close to, the same temperature. In most cases 25 degrees of difference won't matter, and some recipes are flexible enough to cook *mas o menos* 50 degrees, though cooking duration will need tweaking.

- Roast beef and chocolate cookies sharing an oven? You'll find more information on what's best cooked together—and what you may want to avoid—in the sidebar on page 50.

## Plan to Piggyback

Put a new dish into a hot oven as finished ones come out. Much of an oven's heat is lost in the preheating and cooling stages, so this method can save a chunk of fuel; you just need to plan ahead. Use it when simultaneous cooking isn't practical, like when two recipes must cook at drastically different temperatures, or if moisture from one will interfere with the crispiness of another. For instance, as you remove bread or cake or some other dish that needs to cool before eating, put in tonight's dinner, timing it so the dinner is done when everyone's ready to eat. Or vice versa: pop the bread or cake into the oven as you pull out a casserole, potatoes, or roast chicken. Set the timer for the baked goods, enjoy the meal, then retrieve the baked goods when the timer goes off.

## Go Long: Size and Shape Matter

Smaller cooks faster, but did you also know that the greater the surface area, the faster the food will cook? Pick the roast with the most surface area and it will not only cook faster, you'll have more of the deliciously browned outside and ends to fight over. Choose long, thin pieces, for instance, over plump ones, and reduce your cooking time. Same goes for casseroles and baked goods: more surface area means quicker cooking, and if you divide the recipe into two pans instead of one, you'll also shorten the cooking time.

## Kickstart with a Microwave Oven

On the plus side, microwave ovens are more fuel-efficient than conventional ovens. On the down side, they don't heat evenly and don't brown foods well. But you can shave time off conventional baking or reheating by partially cooking the dish (casseroles especially) in the microwave oven, on full or partial power. Finish in a regular or toaster oven for crispness, browning, and complete cook-through.

Finally, at the risk of hearing dedicated bakers utter "Sacre bleu!" let me add this: bakeries are communal by nature: they cook multiple items at the same time. They may not mix savories with sweets (though meat pies and

## Clearing Up Convection: True, Partial, and Faux Convection

Any oven equipped with a fan to circulate air seems to be labeled a convection oven, but there are important differences. Standard, microwave, and countertop ovens (which are slightly larger than toaster ovens) now offer convection options, so research before you buy.

"True convection" ovens feature an added heating element behind a fan on the oven's back wall. (Some professional models include hidden side heating elements as well.) In the most efficient models, the fan recycles the oven's heated air back through the element and out again. Also known as "third element" convection (joining the top and bottom elements), this is the most efficient type of convection oven and the one that cooks most evenly. Wall ovens and ranges feature it more often than do smaller models.

Another type of convection oven circulates the oven's heat with a fan, but without the third heating element. It's more efficient than a traditional oven, though you still get hot and cold spots (meaning you may need to rotate the food to ensure even cooking). You'll find this type built into both large ovens and toaster ovens.

The least efficient models sport an external fan that pulls outside air into the oven, mixing nonheated air with heated air, and defeating any real energy-efficiency gains; avoid this type of "faux" convection. Be prepared to research the model before you buy, as the type of convection may not be obvious.

Also, manufacturer's options can include separate convection settings and special instructions for baking, broiling, and even enhanced browning.

empanadas can be exceptions), but in the Middle Ages, the baker's oven was the town's oven, and households brought their breads and anything else needing oven heat to the baker. It's time we treat our home ovens as communal bakeries, by maximizing their output every time we turn the oven on. This means planning what goes into them simultaneously, piggybacked dishes, shorter cooking times through smaller pans, and adapting seasonal cooking habits accordingly. With practice and planning, energy-conserving baking and roasting can almost be "a piece of cake."

## In Convection We Trust

Convection ovens use about 25 percent less energy than a standard oven. For years I never touched my oven's convection mode. It intimidated me. I was afraid whatever I cooked would turn out *wrong*. Most recipes lack convection cooking instructions, and even though adapting them to convection cooking is simple, I wasn't in tune with energy conservation as an additional motivator.

Looking back, the only thing that turned out wrong was my misguided apprehension. Convection cooking is not as mysterious as it seems—it basically uses a fan to move hot air around—and it can even improve the results, creating lighter baked goods, juicier meats, and tender, caramelized vegetables.

So now I'm a convert: cooking in convection mode is almost a no-brainer, and compared with conventional ovens, happily saves both fuel and time.

### Guidelines for Adapting to Convection Cooking

Here's how convection cooking works: a fan in the oven circulates the hot air around, so the food cooks not just quicker, but more uniformly. Essentially, you bake or roast as you normally would, but set the heat 25 degrees or so lower than normal, or you cook for less time, or do a combination of both.

- If the recipe calls for less than 15 minutes of conventional cooking time: keep the time the same but lower the temperature by 25 degrees F. Check for doneness a few minutes prior to expected finish time. Many vegetable dishes fall into this category.

- If the recipe calls for at least 15 minutes of conventional cooking time: for pastries and baked goods, lower the temperature by 25 degrees F, and be prepared to shorten cooking time by 10 to 25 percent. OR: For meats, roasts, and casseroles, keep the temperature the same and reduce cooking time by 25–30 percent. Check for doneness 5–15 minutes prior to expected finish time.

- Convection time savings increase exponentially: the shorter the conventional cooking time, the less overall time will be saved; and the longer the cooking time, the more likely you'll shave off as much as 35 percent overall cooking time. But regardless of overall savings, convection cooking still out-greens conventional cooking.

- Preheat the oven for most baked goods. For roasts and casseroles, skip preheating and pop the pans into a cold oven.

- Convection ovens excel at cooking multiple items simultaneously. Leave at least 1 to 1-1/2 inches of space between the pans, and above and below them. Generally, you may not need to rotate the pans, as the circulating air should cook them evenly. (But check midway through to see if they're cooking uniformly, and rotate or swap their placement as needed.)

- Heat needs to circulate. Never cover oven racks with foil, especially in convection mode.

- For best browning, use shallow pans and rimless cookie sheets. Cook meats on a roasting rack in a low-sided pan, uncovered (if it browns too quickly, shield with foil).

After you've tried convection cooking on a few recipes, you'll get a feeling for how to adjust your own recipes to get the best results.

## Oven-Free Braises and Stews

Braise or stew entirely over a burner, instead of in an oven, and you'll save a bucket load of fuel.

Most braises and stews start in a pot over a burner anyway, for browning and combining ingredients. Some cooks transfer the pot to a hot oven because the oven continues cooking from all sides, with gentle and consistent heat. But this step isn't necessary.

Modern times offer an assortment of oven-free, mix-and-match methods: cooking on the stove in cast-iron cookware, over an induction burner, in a pressure cooker, or in a slow cooker—all covered in other parts of this book. Briefly speaking:

- If using a heavy pot, like a cast-iron or enameled cast-iron Dutch oven, stir occasionally to prevent burning on the bottom, cover the pot, and keep the heat low, so the contents don't boil but rather stay at or below a simmer. Iron pots are ideal for use on energy-efficient induction burners, too.

- Pressure cookers brilliantly cook braises and stews in minutes, so they consume even less fuel than a cast-iron or heavy pot.

- Slow cookers are even more foolproof. Food won't scorch on the bottom; they can run for hours unattended; and the results are tender and moist. Cooks with a passion for oven-braising may snub "Crock-Pots," but since slow cookers consume about as much energy as a light bulb, it's hard to argue the merits of oven-braising over planet-friendlier alternatives.

If you do braise in the oven, stretch the oven's fuel by multitasking—baking other dishes at the same time. And here's a tip for smaller ovens: flip the lid of a Dutch oven upside-down, so the dome faces inside the pot. Liquid will condense on the top and drop back onto the food (a good thing), and you can roast vegetables, heat breads, or cook other foods separately in the inverted part of the lid.

# Toaster Ovens

Today's better toaster ovens are like the Mini-Me's of conventional ovens. My toaster oven bakes quiche, casseroles and gratins, and cornbread and cakes; it heats and toasts bread, warms enchiladas, broils fish and garlic bread, and works remarkably well, especially for smaller households. Like a compact car, it's a flexible, efficient alternative to a full-size fuel guzzler.

## Toaster Ovens vs. Conventional Ovens

Toaster ovens consume half the energy or less of a full-size oven. But that's just the green beginning. Toaster ovens also *waste* less fuel: both the broiler and oven functions heat up and cool down faster. Because they're smaller, they radiate less heat into the room (hence less impact on room temperature). Their smaller size requires fewer materials to create them and less energy to transport them. For a single item of moderate to small size, they absolutely make more sense. (Some ovens even accommodate a 12-inch pizza.) They can even augment a standard oven during times when you need to cook two dishes at different temperatures. What's not to like?

## Think Twice

If you think you need a double wall oven, stop and consider whether a single oven augmented by a high-end toaster oven might be more practical. It's a greener combination and saves a fistful of dollars. Double wall ovens, if used frequently for cooking multiple dishes simultaneously, like for big families, may be fuel-smart, but cooking a single item at a time in each oven is a hefty waste of fuel.

## Warning

Forget cheap toaster ovens. Invest in the higher-end models with performance that rivals conventional full-size ovens, and consider convection options. Ironically, some models shine in every department except making toast (when compared to a toaster), but that's not how I use my toaster oven anyway. I use it for stuff like garlic bread, baked potatoes, small casseroles, roast chicken, and broiled fish (try that with a standard toaster!). What should you look for in a toaster oven? Cooking and consumer magazines and Web sites regularly test the newest models, so seek out their latest reviews for ratings of:

- Wattage
- Oven capacity
- Maximum-size cookware dimensions
- Actual footprint (not carbon one)
- Convection options
- Racks
- Where the oven is made (how far it's traveled)
- Warranty

## Toaster Oven Cookware

A smaller oven naturally means more compact cooking vessels. This is why I started using glass pie pans for so much of my cooking. They fit in the toaster oven, and because they're glass, they also go into the microwave. You probably already have other pans that fit the space. Smaller baking dishes, bread pans, round cake pans, square cake pans, and tart pans can be used for everything from roasting a small chicken to baking moussaka for four persons. The toaster oven also comes with its own broiling pan (and rack), so an average kitchen should have enough cookware to get started.

# Microwave Ovens

Microwave ovens use 50 to 65 percent less energy than conventional ovens, without heating up the kitchen. Foods can cook or reheat in the same containers used for serving or storage, meaning fewer items to wash and more water and energy savings.

Microwave ovens cook and reheat small amounts of food extremely efficiently. And vegetables steamed in a microwave turn out bright and crisply tender, without needing a pot of water. But large pieces of food, breads, and anything that needs browning are best suited for other modes of cooking, and boiling water in a microwave can take as much time as heating it conventionally. Avoid defrosting foods by microwave; they thaw unevenly or can cook on the edges yet be frozen in the center.

The biggest drawback of microwave ovens is that they don't heat foods evenly and can leave cold spots. The reason: microwaves don't penetrate very deeply, and while certain types of food composition absorb microwaves readily, others are more stubborn. (Microwave ovens generate high-frequency radio waves to penetrate and excite water molecules.) The solution: compensate for uneven cooking by stirring and rearranging food periodically, and by rotating the container on a turntable as it cooks.

Basic microwave ovens are comparable in efficiency to fancier counterparts. Most people (myself included) rely almost entirely on the basic settings: temperature (from high to low) and time (minutes and seconds), as these are the most intuitive. A rotating turntable is an important feature, because it helps food cook more evenly. Bells and whistles like temperature

## Fuel-Free Warming Oven

Need to keep a hot cooked dish warm for a while? Stick it in your microwave oven and close the door. (Don't fire up the microwave unless you want to actually reheat the dish.) The compact size, air-proof seal, and insulating properties of the microwave allow it to hold onto heat very efficiently. I often put hot containers of take-out food, like rotisserie chicken or Chinese cartons, in the microwave for up to half an hour while I get the rest of the meal together.

probes, sensors, and variable power settings can be handy, but may not be important to the way you cook.

### Greener Microwave Guidelines

For safe and efficient cooking, follow these tips, which combine the USDA's guidelines with green practices.

### Use Proper Cookware

Many companies, like Pyrex and Anchor Hocking, make multifunctional glass cookware suitable for microwave and oven cooking, and for storing in the refrigerator (or freezer). Some come with glass lids, others with heavy plastic lids. Glass containers are inherently greener; they're made from natural materials. Borosilicate glass is the toughest; it's the kind used in chemistry labs and also for some cookware.

The USDA conservatively recommends using only cookware specifically made for microwave use, but you can use plenty of your existing glass and ceramic items (no need to buy more stuff). Make sure they have no paints or gilding, and are lead-free. Avoid imported ceramics not clearly marked as lead-free or microwave safe; the glaze likely contains lead. Also, if the plate or container heats up faster than the food, don't use it.

The USDA approves of plastic containers and plastic wraps made for microwave use. Whether recycled or virgin, plastic materials are not the world's greenest choice. Even when labeled microwave-safe, these plastics can melt from the heat of the food, and containers wear down over time. Approved plastics are not supposed to leach toxins, but the potential for long-term effects remains questionable; microwave-safe glass and ceramic materials don't melt in a microwave, so they are my top picks for safe cookware. If you do use plastic containers, follow these USDA cautions: never heat margarine tubs, take-out containers, whipped topping bowls, and other one-time use containers that can warp or melt, possibly causing harmful chemicals to migrate into the food. (Jump to page 147 in Chapter 9 for concerns about BPA, a specific element in some plastics.)

## Brighter Broccoli, Greener Green Beans

How do you keep green vegetables looking pretty? Never cook longer than seven minutes, by *any* cooking method. As the vegetable's cell walls burst, at around seven minutes of heat, a chemical shift occurs, mixing chlorophyll with acids. The result is that drab, ugly hue known as Institution Green (no wonder some kids won't eat their broccoli). But just before this shift happens, air is released, and it clouds the space between the cells, so for a short window of time, the color actually brightens.

Solution: cook smaller pieces, and let the released vapors dissipate and dilute. Microwaves agitate water molecules, and vegetables are mostly water, so despite what most recipes say, *when microwave-steaming vegetables, you don't need to add water*. Water left clinging to the vegetables after rinsing is usually sufficient, and with some vegetables even this isn't necessary. Unless the vegetables are tough or dense, or you're cooking small amounts, don't cover the container, either. The microwave oven cavity acts as a giant steamer, one perfectly large enough to prevent pesky acids from wreaking havoc with the color. Do stir or mix the vegetables several times as they cook, and salt them at the middle or end of cooking, so the salt doesn't leach out too much liquid. To put these tips into practice, try the microwave recipes for asparagus, snowpeas, and corn in Vegetable Sides.

## *Pick the Proper Disposables*

The USDA reports that microwave-safe plastic wraps, wax paper, cooking bags, parchment paper, and white microwave-safe paper towels are safe to use. However, when microwaving, don't let plastic wrap touch the surface of the food. Also, don't use plastic wrap made with PVC and plasticizers; wraps made with LDPE (low density polyethylene) are safer and break down more readily (Natural Value is one brand). To cover a dish without plastic wrap, invert a plate or saucer on top, or even a cotton dishtowel. Never use thin plastic storage bags, brown paper or plastic grocery bags, newspapers, or aluminum foil in the microwave oven. (Again, opt for a glass lid if you can, or a reusable plastic lid that's microwave safe; some companies make universal lids to fit any container size.)

## *Ensure Even Cooking*

To avoid "cold spots" where harmful bacteria can survive, stir or rotate food midway through the microwaving time. A turntable rotates the dish during cooking, but you should still churn the food around if possible, at least once. Moist heat helps cook more evenly and helps destroy harmful bacteria. If covering, leave the lid ajar to let steam escape. Cook large portions of food or pieces of meat on medium power (50 percent) for longer periods, rather than high power for a short time. The center will cook more evenly, and there is less chance of overcooking outer areas.

## *Jumpstart a Casserole*

Shave baking time off a casserole by warming it on medium-high power in a microwave until warm to the touch. Finish the dish in the oven, heating it just until browned on top and cooked to a minimum temperature of 165 degrees F.

When partially cooking food in a microwave oven that will finish cooking by conventional oven, grill, or cooktop (to save time and fuel), transfer it to the other heat source directly; that is, don't store it unless fully cooked.

# Greener Broiling

Broiling is considered upside-down grilling: radiant heat blasts from above rather than below. Steaks, chops, chicken, fish, and vegetables are typically broiled from beginning to end of cooking. But broilers can also act as finish-

## Deeper Debunking: Micro-Managing

A 2003 Spanish study stirred the pot regarding microwave cooking and nutrition. The report noted that more flavonoid antioxidants, which are water-soluble, were lost when broccoli was cooked in a microwave than when cooked in a steamer. But misleading, paraphrased media reports (including ones by notable eco-themed publications) failed to mention a whole bunch of pertinent details. A number of experts have valiantly tried to clear up the confusion.

According to WebMD's Elaine Magee, *"The researchers noted a huge loss of flavonoid antioxidants (about 97% were lost) during the [microwave] cooking process. But the results aren't really useful to those of us who microwave with almost no added water: The researchers used 2/3 cup of water to microwave just 1-1/2 stalks of broccoli. When the broccoli was steamed [on a rack over boiling water] instead, the researchers noted that only 11% of the flavonoids were lost. So the moral of that study is to use as little water as possible when microwaving vegetables."*

Robert Wolke, author and scientist, noted that "because the Spanish study has been misunderstood and misreported, it has triggered undue mistrust of microwaving. But the Spanish research results have nothing to do with the effects of microwaves." Dr. García-Viguera (the report's own author) says that other studies don't show a high loss of nutrients with microwave cooking. For the highest nutrient retention, concludes Wolke, "We should cook our vegetables for as short a time and with as little water as we can get away with"—something microwave ovens do extraordinarily well. (Read Wolke's complete analysis at http://www.heraldextra.com/content/view/168705.)

ers: browning the surface of casseroles, gratins, and proteins that may need an extra bump to give them that sexy golden-brown appeal.

Broiling cooks food hot and fast, for a lovely exterior sear and crispiness. The downside is that it takes a lot of fuel to get the broiler hot enough to actually broil. Except for toaster ovens, the minimum time needed to preheat a broiler is 10 minutes, and 20–30 minutes may be necessary in some cases.

Some cooks even heat the oven to 550 degrees first, before jumping to the broiler setting. And there's no green way around it: preheating the broiler is a must; cold-start broiling just doesn't work.

In standard ovens the electric broiler element is at the top, and in gas units it's either on the oven's ceiling or in a drawer below the oven. Gas broilers tend to get hotter and be more effective than electric ones, but both can suffer from poor design: the element typically doesn't cover the broiling surface adequately. Plus, broiler elements cycle on and off, to regulate heat and prevent overheating, but this can be counterproductive. Instead of broiling hot and fast with radiant heat, you end up in a kind of wishy-washy, baking-broiling mode. (Most home broilers cook at 10,000 BTUs, but you really want closer to 17,000 BTUs for true broiling.)

So how can you get the ideal broiling effect without using as much fuel? Consider these strategies:

## Use a Toaster Oven

Broil in a toaster oven. It takes less time to bring the element and surrounding air up to high temperature. The food can sit closer to the broiling element, and the heat all the way around the food gets hot enough to speed up browning and searing. Plus, if you want to zap a golden finish on a skillet-cooked dish, like an omelet or chicken pieces, run it under a toaster oven's broiler for just a few minutes.

## Dry the Surface

When broiled, moist surfaces steam, but dry surfaces brown. So blot up moisture and excess marinade before cooking. (Dry rubs are fine to leave on.)

## Oil the Surface

A light coating of fat conducts heat, speeding up browning. Rub steaks, skinless chicken, and chops with a thin coating of olive oil. When preparing peppers for peeling, lightly rub them with vegetable oil; they'll blacken quicker.

## Cook Close

Cookbooks often say to broil as much as 6 inches from the heat (meaning the food surface's distance from the element, not the pan's), but if the pieces aren't very thick and there's not a lot of fat, broil closer: as close as 3 inches from the heat for some foods. Fatty foods like chicken skin will splatter and may burn if too close to the heat; go for leaner cuts and thinner cuts if cooking close. If you can't raise the oven's rack higher, slide an inverted pizza pan or baking sheet under the broiler pan to raise it up a skosh.

## Heat the Pan

Stick the empty pan in the broiler while it heats up. Traditionally, you cook meats and poultry on a broiler rack over a pan with low sides to catch the juices. Cookbook author Mark Bittman recommends ditching the rack and heating up a cast-iron (plain or enameled) skillet; add the food to the hot pan, which sears the underside, and let the broiler element finish the top side, without flipping the food. This works fine if you don't have a lot of juices escaping (otherwise use the rack method).

## Pan-Fry or Pan Grill Instead

A heavy skillet or grill pan on the stove can create the same seared, crisp exterior on meats, fish, and poultry as a broiler, and it uses far less fuel. Let the pan get good and hot, and the ridges of a grill pan will lift the food so it doesn't boil in its own juices. (As with broiling, dry the food surface, and lightly coat the surface with oil to help conduct the heat.)

*One last tip:* If you do broil on a rack, capture any cooking juices and save them for stock or soup or adding to rice or a sauce.

# 25 Green Tips for Blue Ovens and Broilers

1. Preheat an oven only if necessary. Baked goods, roasted meats, and poultry are more likely to need initial hot cooking, but most casseroles can go directly into a cold oven.

2.  Don't start preheating earlier than necessary (no need to waste fuel). Ten minutes should be ample for most recipes, and fifteen minutes is good for very high heat recipes.

3.  Don't open the oven door unless necessary. The temperature drops about 25 degrees every time you do.

4.  Adjust your oven racks to the desired positions before heating the oven.

5.  Use corded thermometers to monitor the internal temperatures while food cooks, so you don't have to open the door.

6.  Don't overcook. This may sound obvious, but overcooking happens. Set a timer and use a thermometer.

7.  Clean your oven window to monitor cooking progress. Like the windshield of a car, if you can't see through the glass you can't tell where you're at. Scrape away burned-on gunk with a razor blade.

8.  Line your oven floor with special oven mats, and clean those as needed instead of turning on the self-cleaning feature.

9.  Cook with the convection mode if you have one.

10. Keep air flowing: don't cover racks with foil, and don't block heat vents on the oven floor with foil, either.

11. Stagger dishes and leave at least an inch of space between them so the heat circulates more evenly.

12. If you have more than one oven, opt for the smallest one that will do the job, like a toaster oven. Try to adapt favorite recipes to cooking in a smaller oven, too; sometimes all it takes is a different pan size or shape.

13. Bake or roast more than one item simultaneously. Ask yourself: What else can I cook at the same time?

14. Or, piggyback. Since the oven will be hot, can you pop something else into it when the first dish is done, and make use of the existing heat? (Getting an oven up to speed uses a good deal of fuel.)

15. Cook on the stovetop instead of the oven or broiler whenever possible. Braise and stew over a burner, at a fraction of the fuel used by an oven. Pan-fry in a grill pan instead of broiling.

16. When using glass and ceramic cookware, you can generally shorten cooking time or lower the temperature by 25 degrees F.

17. Use the self-cleaning feature only when necessary. Wipe up spills before they bake on.

18. Turn on the self-cleaning feature right after you've used the oven, to take advantage of the existing heat.

19. Make sure the oven door's seal is in good shape, with no tears or gaps; replace if necessary (contact the manufacturer or an appliance repair store).

20. Cook in smaller sizes to shorten oven time: make two smaller pans of lasagna instead of one large one, or cook muffins instead of a single loaf.

21. Cook once, enjoy twice: double the recipe, morph or freeze your leftovers, and freeze a bunch of baked goods to maximize fuel use and minimize the cook's time. It can take far less fuel to reheat a meal than to cook a new one from scratch.

22. Tweak recipes to cook them simultaneously. When the cooking temperatures range from within 50 to 75 degrees of each other, split the difference and cook them all at the midrange setting. For instance, if one recipe calls for 350 degrees F, another for 375 degrees F, and a third for 400 degrees F, set the oven for 375 degrees F. Shave a few minutes off the cook time for the lowest temperature dish, and cook the higher temperature dish a little bit longer.

23. Turn off the heat early: let food finish cooking in the oven's residual heat. And once food leaves the oven, let it rest a few minutes for the internal heat to continue its "carry-over cooking."

24. Bring foods to room temperature before placing in the oven. (And be sure to thaw foods thoroughly before cooking as well.)

25. Don't use your oven at all. Opt for an alternative preparation method, like no-cooking, slow-cooking, cooktop cooking, or another more fuel-efficient alternative.

Finally, consider the cookware: see Chapter 6 to discover how cookware affects fuel efficiency.

# FOUR

# Green Flames:
# Cooktop Cooking

*There is no one way to get what you want unless it is to remain open. Keep guessing. There are as many ways as you can think of. The best one is the one that works for you. NOW. Maybe it never worked before. No one has ever fried an egg without turning on the gas, but maybe this time if you look that egg straight in the eye and say "FRY," it will. Sure it will be the first time, but a whole new world will open up and that's a gas. So if you feel like trying something unheard of, do it! ... Try it your-self ... your own way. NOW, LET'S EAT! ! ! ! !*

—Alice's Restaurant Cookbook

The simplest way to shrink a cookprint is to reach for cooktop recipes first, rather than oven ones. What's so hard about that? It's a good starting position, and cooking over a flame is the world's most common method, especially where fuels are scarce. Cooktops can even replicate an oven's baking and broiling effects, without the same overwhelming energy loss. So if you already cook mostly *on* the stove instead of *in* it, give yourself a green star.

Going green is all about choices, and with stovetop cooking, you've got plenty of tasty green options to consider. As mentioned earlier, all forms of cooking boil down to one critical feature: controlling water. On the stovetop, you manipulate water every time you boil, steam, blanch, and braise—and also when you fry, sauté, and stir-fry.

For these reasons, this chapter explains how to cook "with a green flame"—traditional methods requiring less fuel, radical makeovers of these same methods, and a few surprising "plug-ins."

## How Cooktop Cooking Works

A cooktop and a stovetop are functionally the same thing: the cooktop is an independent appliance whereas the stovetop is attached to an oven as part of the range.

In both cases, the heat source comes from below. The challenge lies in channeling that energy from below into the food. The heat must first penetrate a vessel of some kind (a pot or pan). But it's not done yet. The heat then needs to jump through yet another layer of conductive "stuff," which can be water, fat, or air, before having any effect on the food. (Even in dry-pan cooking, food has a surface layer of fat or water that affects the way heat penetrates.) Finally, there's the food itself, and each form of food "material" responds to heat (or cold) uniquely.

But cooking isn't about fuel efficiency alone, it's about taste. You could eat pot roast or you could eat shoe leather. Converting a tough piece of meat into a tender and flavorful dish requires adequate fuel, water, and time; pan-frying uses less fuel, no water, and little time, but you'll end up with more of a football than food.

So with cooktops, the cookprint you create relies on manipulating three sets of characteristics: the vessel material, the food material, and the conductive "stuff." In the new green basics of cooking, the trick is to mix and match for optimal fuel efficiency—while still creating delicious, nutritious meals everyone will enjoy.

Immersing foods in hot water or surrounding them in steam have pros and cons. On the one hand, the basic process requires a lot of heat just to bring water to a boil. Additionally, you consume not just fuel but also *water* as the heat conductor. In drought-plagued, water-restricted areas, cooking with boiling water may be more precious than oven roasting. But even where water remains plentiful, there's good reason not to waste it. So what follows is a new way to look at how you cook with water: to conserve both the water you use and the fuel to heat it.

## Passive Boiling

*The Zen of the Green Flame = Do nothing.*
*You've created the energy. Now let it work on its own.*

Cooks are rethinking the ways they cook with water. They've discovered that certain foods turn out best not by boiling them until done, but by cooking them in water as it heats up, or as it cools down, or both. I call this green-flame strategy *passive boiling*. For instance, you can turn off the heat when the water boils (or even before) and let potatoes or pasta "cook" in the heated water as it cools.

Besides saving fuel, passive boiling cooks foods more gently than the harsh, raging bubbles of a fast boil. Because the water temperature lowers toward the end of the cooking period, there's less chance to overcook by detonating the vegetable cells into mush or causing the proteins in meats and seafood to toughen.

Boiled water takes a long time to cool, which is what makes passive boiling work. The passive boiling method brings cold water (with food submerged) to its hottest possible stage (boiling), but only briefly, then gently allows the slowly cooling water to penetrate deep into the center of the food. In some cases, foods like pasta may need to be boiled for a few minutes at first, though far less than the traditional length of time, and the rest of the "cooking" is done entirely fuel-free. In other cases, the water doesn't even need to reach boiling, a technique that saves a surprising amount of fuel just by knocking off one degree of temperature, before it undergoes "phase change."

## Boiling Strategies

Ever wonder why a watched pot never seems to boil? Water has a high heat capacity, meaning it requires a hefty amount of energy to reach its boiling point of 212 degrees F (100 degrees C; at sea level). Most of this energy goes into heating the water to one degree below the boiling point, which it reaches fairly quickly. But that final one degree jump, from pregnant possibility to full-steam boiling, needs half that same amount of energy as it took to get there in the first place. Consequently, the wait for getting to a true boil from what looks like a ready-to-boil pot (phase change) really can seem interminable. What can you do to hasten boiling?

### Cover the Pot

A lid traps heat and water being lost to evaporation, so a covered pot really does boil faster than an uncovered one.

## Turn Down the Boil

Once the pot boils, the temperature remains constant, which is its main asset for delivering consistent results. But a rapidly boiling pot of water is the exact same temperature as a gently boiling pot. Food cooks in the same amount of time, whether the water is boiling raucously or sedately. Churning water helps separate strands of spaghetti and disburse the starches, but after the first few minutes, even pasta doesn't need to boil (as the recipe on page 199 shows).

## Savor the Simmer

Many foods cook just fine in the sub-boiling zone (210–211 degrees F). Vegetables blanch well, delicate teas taste better because there's more oxygen in the water, and even surface bacteria die within fifteen seconds or less. Poached salmon wouldn't be the same if cooked in true boiling water. (Cutting-edge cooks may want to explore *sous vide* cooking, a method of long, slow cooking in water well below the boiling point.)

## Turn off the Heat

Practice "passive boiling" by turning off the flame (either immediately upon boiling or after a few minutes after) and letting the energy from the boiled water cook the food. The next section charts the many ways to make use of passive boiling.

### Passive Boiling Tips

Long before I knew about greenhouse gases, I'd read about how Chinese cooks practiced passive boiling (as in the juicy Ginger Chicken and Broth, Passively Poached on page 177). In the previous century, American recipes for deviled eggs began to use the term *hard cooked* instead of *hard boiled*— because heating eggs without boiling them yields perfect, tender yolks.

What can you put to a passive boil? Just about anything that needs some time in boiling water can adapt to this greener method. As mentioned, spaghetti can cook passively, and for even more energy savings, some Asian pastas need no boiling at all; they soak in tap water until soft (perhaps these were developed because fuel was scarce?).

Keep in mind that passive boiling can work in two ways. Some foods start in cold water, heat is added until the water boils, and then the heat is turned

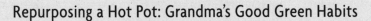

### Repurposing a Hot Pot: Grandma's Good Green Habits

Once boiled, water takes an impressively long time to cool down. Grandma's hot water bottle warmed people, clothing, towels, beds, and even pets. If you're not turning hot cooking water into soup or stock, put the energy released from it to good use. You can scoop out boiled pasta, cover the pot, and use it as a passive thermal heater (for plants near a window, or a pet sleeping area). A pot of passively boiled chicken can rest anywhere, too, while it slow cooks, flame-free. Set the pot on a stack of towels for a cozy post-bath treat. If you save jars, soak them in leftover hot water to remove the labels. Pour a cooled pot of unsalted cooking water into a watering can for your house plants, garden, or as a drink for thirsty wildlife.

off (either right at or after a few minutes of boiling). Other foods start in boiling water, then cook for at least part of the time with the fuel turned off. Try putting a green flame to common ingredients, by following the chart on page 72 (times are approximate)—and use it as a guide for other similar foods.

## Passive Blanching

In blanching, also known as parboiling, food is immersed in boiling water, but just until it's partially cooked or, for vegetables, crisply tender. Blanched foods are typically plunged into boiling water, scooped out, and shocked in a bath of ice water to arrest the cooking process. If you're going to expend that much fuel to bring a pot of water to a boil, stretch that fuel by blanching several foods in succession. Repurpose the blanching and the ice water in soups, stocks, or beans; for nonedible usage, soak jars in hot water to remove the labels, and pour the cooled water onto plants.

Why blanch? Because it takes the rawness out of vegetables, shrimp, and other foods destined to finish cooking by another method (like baking in a casserole or stir-frying) or that should be undercooked, to retain some

| TO COOK | START BY SUBMERGING IN | ACTIVE BOILING TIME | PASSIVE TIME IN HOT WATER |
|---|---|---|---|
| Large egg | Cold water | None; turn off heat as soon as water comes to boil | 12 minutes for hard cooked (plunge into chilled water to stop cooking); 5 minutes for soft cooked |
| Whole russet potatoes | Cold salted water | 0–10 minutes, depending on size; can be boiled with other foods simultaneously | 20–30 minutes, covered; pierce with a skewer to check for doneness |
| Small red potatoes | Cold salted water | None; but can boil while blanching other vegetables in same pot | 10–15 minutes, covered |
| Corn on the cob | Boiling salted water | None | 6 minutes covered, or until tender |
| Green beans (1-1/4 inch pieces) – some active boiling | Boiling salted water | 2 minutes | 3–4 minutes to fully cook |
| Green beans (1-1/4 inch pieces) – no active boiling | Boiling salted water | None | 5 minutes to blanch |
| Zucchini (1/4-inch slices) | Boiling salted water | None | 2 minutes to blanch |
| Whole chicken | Cold salted water with seasonings, aromatics | 15 minutes at a simmer, after boiling is reached | 1 hour, covered |
| Chicken breast halves, with bone and skin | Cold salted water with seasonings, aromatics | 5 minutes at a simmer, after boiling is reached | 20–30 minutes, covered |
| Shrimp | Boiling salted water | None | 1–5 minutes, depending on size |
| Spaghetti and similar pasta | Boiling salted water | 2 minutes | 8 minutes, or until al dente |
| Rice | Double its own volume of cold water | 10 minutes at a simmer, after boiling is reached | 30–60 minutes, until rice absorbs water and is tender |

crunch, as part of a salad for instance. Blanching essentially flash-boils food, and the process can help retard deterioration of proteins (like shrimp) and vegetables.

Because water takes several seconds to return to a boil after the food is added, and since most vegetables and shrimp cook so quickly, does the water really need to return to a boil at all? In most cases, I discovered that "passive blanching" works just fine—pour boiling water *over* the vegetables, then let them set until tender—and boiling the water in an electric kettle can save fuel. For best results, cut the vegetables into smallish pieces and arrange them in a single layer in a shallow glass baking dish. Put the method into action with Teapot Blanched Green Beans on page 74.

## Passive Blanching Tips

For traditional blanching, you need a pot and a big bowl for shocking. You can make the process greener with a few handy items and techniques (including saving the water for plants or other uses).

### Right-Size Pot and Lid

As discussed, water boils faster with a lid on the pot. Pick the right size pot for the job. You'll need a large pot (5 quarts or so) for blanching multiple items sequentially or for large volumes of food, but you can scale down to a 2-quart pot for a half-pound or so of fresh vegetables.

### Spiders, Strainers, and Slotted Spoons

A Chinese spider is a wire-mesh skimmer with a bamboo handle, and they come in various sizes (5–6 inch diameter is handy for blanching and frying). With a spider, you can use the same water to cook multiple foods in stages, scooping them out of the pot rather than draining in a colander. A slotted spoon or a wire-mesh strainer with a handle will also work.

### Shocking in an Ice-Free Bath, 1.0

To conserve water when shocking vegetables, keep reusable ice packs (the kind meant for ice chests) in the freezer and a jug of cold water in the fridge. Drop the packs into a bowl and cover with chilled water for shocking. Rinse, dry, and return the ice packs to the freezer when done.

## Teapot Blanched Green Beans

Traditional chefs may flinch, but by tweaking conventional blanching methods, you can stretch fuel and water, and still get perfect results. This method to passively blanch green beans also works for other vegetables. An electric tea kettle boils most efficiently, but any source of boiled water will do. Trim 1/2 to 1 pound green beans into 1-inch lengths. Scatter them in a shallow glass baking dish, so they're only 1 to 2 layers deep. Pour in enough boiled water to cover (2–4 cups should do). Then let them passively "blanch" for 3–5 minutes, or just until tender with a crisp bite (older and larger beans take longer). Scoop the beans into a bowl of ice water, or drain in a colander under cold water. After draining, use chilled for salads, reheated with butter, in a casserole, soup, stew, or as you wish. Beans may be refrigerated up to four days. (Apply this method to asparagus, carrots, and other vegetables; the denser the texture, the smaller the pieces should be.)

### Shocking in an Ice-Free Bath, 2.0

Drain the veggies in a colander, then run cold tap water over them, mixing them up so the water cools them quickly and evenly. This works best for small amounts or pieces that aren't very large or thick.

## Reasons to Blanch and Shock

*Stop the aging process, stretch the freshness.* For anyone who's ever over-bought at the farmers market, blanching comes to the rescue. If your snow peas, broccoli, or other fresh veggies start to lose their freshness faster than you'll be able to use them, blanch them until crisply tender. Shock them in an icy bath, drain and let dry, then pack them loosely in a container, seal, and refrigerate (if they're still damp, wrap in a cloth or paper towel before sealing). They'll keep their fresh flavor and color another two to four days.

*Save time.* Blanched veggies instead of raw ones can reduce the overall cooking time when added to other dishes, like casseroles. Better yet, having blanched veggies on hand can be a time-saver strategy for meals later in the week, too. Toss them into salads, stir-fries, soups, casseroles, or noodle dishes, or reheat in the microwave with butter and garlic as a stand-alone side dish.

*Brighten the color of vegetables.* Blanching makes greens greener, reds redder, and other colors brighter by releasing the air between the cells.

*Crisply tenderize.* Blanching softens the fibrous texture of broccoli, carrots, and string beans but still keeps the crispness.

## Steaming

Steaming is a marvel of physics and a gem of a cooking method for most vegetables, seafood, poultry, and tender meat cuts. Steamed foods also retain more nutrients than boiled ones, and with less of their essentials washed down the drain, they tend to taste more purely of themselves.

**Steaming takes longer than boiling.** So why use it? Steam is the same temperature as the boiling water used to create it. But how the process works makes all the difference in the results. Steaming uses a fraction of the water that boiling uses (so less fuel is needed to get it to the boiling stage), and though steaming takes longer to cook foods, it's typically only 10–30 percent longer than if the food was submerged in a big, hot boiling bath. Compared to steaming, boiling is harsher on foods and strips them of more nutrients. Plus, boiled foods need to be submerged in water, which means more water used, and consequently more fuel is needed to bring that larger amount of water to a boil.

What else makes steaming different from boiling? Water as a liquid has a higher heat capacity than water as a gas (aka steam). In plain lingo, boiling water can hold more heat and delivers it more directly than steam does. But when water jumps past boiling to 213 degrees F and becomes steam, a kind of wild, thrashing energy wraps itself around the food, and very quickly, very efficiently, releases itself into the food.

How does this happen? When steam comes into contact with a cooler surface (the food), it condenses. As this energy (steam) changes from a gas to a liquid, it transfers enormous energy (about 500 times as much energy as would be released by cooling the same weight of water by a single degree). That's how the process of steam cooking occurs, and why it's strong enough to power a locomotive engine.

## Assembly-Line Blanching: Try This at Home

*A hot pot of water? Oh, the things you can do! Here's a sequence of foods I put through the blanching/shocking process recently, and how I later served each one. The whole process took less than twenty minutes of active fuel, including bringing water to a boil, and we ate well for a week.*

To set up an assembly line, it helps to have extra ice on hand (and/or ice packs), an extra colander for draining, and some storage containers to dump each food into after the final stage. Start the blanching process with the most tender vegetables and move up sequentially to the starches or proteins, scooping each new food up with a spider and dumping it into a large pot of chilled water. After the first batch cools, I spider-scoop it into a colander to drain. Meanwhile, I blanch the next set of vegetables (or other food) and continue the dance of blanching, chilling, and draining until every batch is done.

- Whole trimmed snow peas, served three ways: first in a stir-fry, next reheated with butter and garlic as a side dish; leftovers from these dishes went into fried rice.
- Broccoli (in half-inch pieces or small florets): added to the fried rice, a vegetable salad, and a bit tossed into a cheesy macaroni; I could also have made a potato-broccoli soup with them.
- Egg noodles, boiled until al dente: half were served that night as a bed for pan-braised chicken; the rest later became a stand-alone side dish tossed with garlic butter; leftovers from these meals were tossed into a soup for lunch.
- Whole potatoes (medium-size), plunged into the boiling water, then covered with a lid and the heat turned off; they were tender after thirty minutes: I scooped out the flesh, cooled it, and refrigerated it for the basis of a hash. (If I'd been thinking about it at the time, I could have added some eggs to passively hard cook with the potatoes, for potato salad.)

When I was finished blanching food, I poured the cooled cooking water into my garden, to feed and nourish the plants. And I set the ice-bath water out for the toads, deer, birds, and other wildlife, especially since we're in another rough, tough drought.

## Green-Steaming Strategies

Whether you use a single-level steamer or multiple tiers, maximize your steam energy with these tactics:

- Set the water level about 1 inch below the lowest food tier; you want the food to steam, not boil.

- Allow enough space for the steam to circulate, especially with multiple tiers. The more contact from steam, the faster and more evenly foods will cook. If you set foods on a plate or solid surface (like parchment paper or foil), make sure there's enough room for steam to come up around the sides.

- Use only enough water as needed to generate adequate steam, without boiling dry. An inch of water over medium-high heat typically lasts 15–30 minutes (depending on the pot size). How can you tell if there's still water in the pot? Drop in a few marbles, a dime or any heat-proof object that will clatter while the water boils. If the noise stops, you've run out of water, or you're not making steam.

- Keep your water boiling to produce steam, but don't waste fuel. If your pot looks like a locomotive, with steam bursting out from under the lid or the tiers, turn down the heat.

- Make sure the lid fits tightly, to prevent steam from escaping. Some cultures seal steamers with dough, wrapped like a ribbon around the rim. For more convenience, wrap a damp cloth around the rim of the pot, or drape it flat across the pot, and set the lid so it fits snugly on top. Ditto for any tiers: they should also fit snugly.

- Cook in multiple layers. To stretch your steaming fuel and water, steam several foods at once. Place faster-cooking foods in the upper tiers. Foods in the lower tiers, closer to the steam source, cook more quickly than those in the upper tiers.

- Save the steaming water for soups, stocks, sauces, grains, beans, or other uses.

  *Warning:* steam burns—badly. Always open the pot lid away from you and protect your flesh with sleeves and mitts.

## Ten Ways to Stretch a Pot of (Boiled or Boiling) Water

*Stagger ingredients or cook them simultaneously, and repurpose the water as part of another dish. Use full boiling, passive boiling, or combo-boiling methods.*

1. Cook regular or sweet potatoes: start in cold water, then blanch green beans or other vegetables when the water reaches boiling. Scoop out green beans and let potatoes passively cook.

2. Blanch and shock green beans, broccoli, and other vegetables in stages; refrigerate for use throughout the week.

3. Peel tomatoes, pearl onions, or peaches: Carve an X in the base and plunge in water a few seconds. Peels will slip right off.

4. Mellow garlic cloves. Drop in a handful of unpeeled garlic gloves. After a few seconds, scoop out. Peels will slip off and garlic will be mellow and mild enough to eat whole. Brown blanched garlic cloves in butter or oil and serve as flavorful garnishes, over vegetables, fish, chicken, meats, or in pastas.

5. Scoop out a cup of water for tea. Before adding foods to the pot, ladle out some water for a nice cup of tea.

6. Or, make a big pot of smoky tea and drop in vegetables like eggplant (see recipe page 228), or Asian noodles, or wontons, or make Chinese marbled tea eggs.

7. Pour the water over couscous or bulgur wheat to cook. Turn regular lasagna into no-boil noodles by soaking in boiled water, then bake as in the recipe on page 203.

8. Cook eggs: hard cook, soft cook, and in-between.

9. Cook shrimp in just a few seconds (when the shrimp start to curl, they're done). Flavor the water first by cooking/blanching vegetables in it, and adding seasonings, if desired. After cooking the shrimp, use the water to cook rice or pasta.

10. Parboil ribs before grilling to tenderize the meat and reduce grilling time. (Blanch meats and proteins after all vegetables or fruits are blanched, at the end of a sequence.)

Steaming gets even more fuel-efficient with these methods:

- Pressure cooking. Pressure cookers provide a kind of fuel-efficient dream steam, with more nutrient retention, faster cooking time, reduced fuel and water consumption, and tastier results.

- Microwave steaming. Steaming is one of a microwave oven's best skills. Microwaves jiggle the water molecules into heat and steam so efficiently that foods cook quickly in little or no water (see page 57).

## Braising and Stewing

Braising and stewing are by definition low-heat, long cooking methods and should never be rushed. That doesn't mean they can't be made more energy-efficient, though some tactics may send Larousse spinning in his grave. (Coq au vin, beef Burgundy, goulash, gumbo, chili, birria, rogon josh, and fricassee are popular examples of braises and stews.)

Typically intended to tenderize tough meat, braising starts with a large hunk of meat, like a pot roast, while stewing relies on smaller chunks (meatless stews and braises use the same method, so if you're vegetarian or vegan, please follow along).

Braises and stews first brown the food in hot fat on the stovetop (though some dishes skip the browning step). Then a liquid is added and the whole thing simmers covered, low and slow, until the meat's fibers (or the vegetable's) break down and morph from rough and tough to tasty and tender. Braises typically call for a small amount of liquid (just enough to come partially up the sides of the food), while stews can use a little liquid or enough to cover the food completely. Layers of deep flavors develop from the additions of aromatic vegetables, herbs, and spices.

You can braise on the cooktop or in the oven. Mastering the stovetop method makes better green sense. Most chefs prefer the oven method, moving the vessel from stovetop to oven, so heat penetrates gently from all sides. This process can take hours, and as we all know, the oven's an inefficient energy hog that spews most of its heat to every place *but* the food.

### A Greener Cookprint with Cooktop Braising

Most civilizations have been braising and stewing low and slow *over* a heat source for centuries. As I researched the origins of the world's most popular stews and braises, I rarely found that these dishes were cooked in ovens; most of the time, the directions called for stovetop heat.

Only where ovens are widespread and fuel is plentiful do we see single braising vessels surrounded by oven heat, or where the method taps into the same source as the home's heat: set inside a masonry or clay hearth, or nestled in a wood-burning stove, for instance.

Stovetop braising already consumes less energy than firing up the oven, but to produce the best and most efficient braise possible, follow these tips:

- Monitor and stir the braise to prevent scorching. Keep the heat low.

- After browning and adding the liquid, keep the liquid from boiling; it should be at a simmer or just below.

- Use a heavy pot with a heavy lid, one that will conduct heat from the sides and top as well as the bottom. A cast-iron Dutch oven (with or without enamel coating) or one with aluminum-clad bottom and sides are good picks. Heavy cast-iron cookware takes a while to heat up, but once it's hot, it stays hot and distributes the heat gently and evenly.

- Go for a snug fit. Pick shallow over deep. You want a fairly snug-fitting vessel, so there's less air in the chamber, but also one that allows for moist heat to circulate.

- Choose pots with lids that allow the liquid to condense and return moisture to the center of the pot, basting the braise as it cooks. Some lids have spikes or ridges; others are slightly convex in shape.

- To use less liquid, rest a round of parchment paper directly over the top of the braise. The flavors will be more concentrated, and the heat will be used more efficiently. (This is a classic French technique, though few recipes today mention it.)

- Stretch your stew or braise by serving it with grains, pastas, or breads. Better yet, dollop some dumpling dough on a stew as it cooks, to end up with a true one-pot meal.

### Add a Lid to Go from Fry to Braise

After browning foods in a skillet or pot, adding a lid changes the cooking method by trapping steam. Want a crispy exterior? Don't add a lid. But if you need the interior to cook without overcooking the exterior, a lid will trap the heat and vapor and turn your vessel into a mini-oven; the green advantage is that foods cook quicker than in a traditional oven, so you're expending less fuel and less time. Adding a lid braises foods, and in stir-frying, it completes the cooking process with just enough vapor to gently cook the stir-fried pieces all the way through.

## Greener Alternatives for Braising and Stewing

Consider these nonconventional methods for lip-smacking, fuel-efficient results. Other chapters in this book cover each item in detail:

*Plug in a slow-cooker or Crock-Pot.* These energy-efficient machines are brilliant for braising and stewing (and they also cook in other ways). You may need to brown the food in a skillet first, then transfer it to the crock, but there's good news: some models now come with a removable insert that allows you to brown foods in it, directly over a burner.

*Practice passive slow-cooking:* Turn off the slow cooker for the last hour of cooking and let the retained heat continue to cook the food. Set a folded towel over the lid to keep more heat in.

*Fire up a pressure cooker.* It's a brilliant fuel-saving device, and you can brown directly in it, before putting the lid on for pressure braising or stewing.

Haybox cooking, thermal cookware, and solar cookers also work low-energy magic, especially with braises and stews, and they're discussed in other pages of this book.

# FIVE

# Frying: Fast and Fuel-Efficient

*A covered pan of water will take more than twice as long as a pan of oil to heat up to a given temperature; and conversely, it will hold that temperature longer after the heat is removed.*
—*Harold McGee,* On Food and Cooking

Let's face it: life without fried foods would be no fun. Globally, we'd have no egg rolls, taquitos, tempura, plantain chips, or fried chicken. The good news is that frying methods (from stir-fry to deep-fry) can be more fuel-efficient than boiling or baking, though they're not without some green drawbacks.

To understand frying, you have to know about fats. Fats conduct heat amazingly well, far better than water (which is in turn better than air). Vegetable oil, for instance, has about half the heat capacity of water, so it heats up quicker and can be raised to higher temperatures.

But in the never-ending trade-off of greener cookprints, most cooking fats are oils pressed from growing plants, nuts, or seeds. They're renewable, but they still require resources and processing to create them. Higher up the food chain, fats also come from animals (as in pork lard, chicken schmaltz, or butter from dairy cows). Fats cook quickly, but they can also release particles into the air, and the need to dispose of spent grease properly drains a bit off their lean, green profile.

# Greener Frying Strategies

The term *frying* collectively includes stir-frying, sautéing, shallow frying, and deep-fat frying. Some frying uses no added fat (relying on the natural fat of the food itself). Not to be confusing, but the term *pan frying* can mean shallow frying or sautéing, and stir-fries can be fried in a pan, too.

Moving from brightest green to the paler green methods, here's a rundown of frying strategies, all of which use less fuel than an oven.

# Stir-frying

If more people stir-fried, the world would be a greener place.

A wok-cooked stir-fry wins at fuel-efficient cooking, beating out boiling as well as all other frying methods.

In fact, most global cuisines make their own style of stir-fries, so the method goes far beyond Asian foods. Although this ancient Chinese method doesn't demand a wok (a skillet works fine), a wok's unique concave shape makes the most efficient use of the flame. Essentially, you need really high heat, a pan that conducts heat well (and withstands high temperatures), very little oil, oil with a high smoking point, and sometimes a lid (often with added liquid, to finish cooking). One caveat: woks were designed to sit on a ring over a true flame; today's electric cooktops have led to flat-based woks, which are still highly efficient. Because of the high heat needed for stir-fries, avoid nonstick-coated pans.

Most of the energy spent in stir-fries comes from you, the cook: food must be cut into small enough pieces to cook quickly and evenly (but there are shortcuts to this, mentioned below). After you've done the chopping, a stir-fried meal can take less than five minutes of fuel to actually cook, and as little as a tablespoon or so of oil.

## Getting into Global Stir-Fries

If stir-frying is so good, why don't more people do it?

I believe that for most cooks, the biggest obstacles lie in the chopping, long ingredient lists, a need for specialty condiments and sauces (like bean paste or oyster sauce), and when improvising, the brain-stress of deciding what flavor combinations go together. Plus, most folks assume stir-fries have to be Asian. Wrong. Toss all of these preconceptions aside to open up a bigger world of energy-efficient, wok-style cooking.

## Chop Shop

If chopping foods is your idea of drudgery, pick ingredients that need little or no chopping, like shrimp, shelled nuts, snow peas, or grape tomatoes. Also, chop a little more than you need every time you chop, and save the extra for a stir-fry later in the week (previously blanched vegetables also increase the inventory; see page 76). If you've been putting your fridge and freezer to good use, you'll make a fabulous dinner in minutes. With a little forethought while prepping Monday and Tuesday meals, by Wednesday you can end up with precut, ready-to-use vegetables (green onions, carrots, green beans, broccoli); proteins (meat, poultry, sausages, bacon, shrimp, and tofu); toasted almonds, pine nuts, or other nuts; black beans or chickpeas from the freezer; and leftover rice or grains as a side dish or mix-in.

## Size Matters

But not as much as you may think. As long as everything's about the same size and is small enough to cook fairly quickly, the main ingredients don't have to be bite-size—they can be as small as peas or as plump as jalapeños, fine as matchsticks or as thick as your thumb. Smaller ingredients cook quicker, but even larger stir-fry ingredients cook more efficiently than in other methods.

## Anything Goes

Stir-fry anything, not just Asian foods. Try Mexican taco fillings of zucchini, poblano strips, and onion with cumin and chile powder (chicken, beef, or seafood optional). Portuguese linguiça or Spanish chorizo sausage with sweet peppers, corn, and smoked paprika. French-inspired wild mushrooms with turkey, cream, and thyme. Indian potatoes with curry powder, chickpeas, and coconut milk. Or eggplant, fennel, and tomatoes with paper-thin Parmesan shavings.

## Focus on Flavors

Save time by limiting the number of ingredients to just a few flavor-intense ones. One or two vegetables or a protein is sufficient, along with aromatics like onion, garlic, and/or ginger, and a blast of one or two spices or seasoning blends. A final touch of fresh herbs, like basil, or a sprinkling of pretoasted pepitas or sesame seeds, easily rounds out a dish with a manageable number of ingredients.

## Sautéing

Sautéing lightly coats the pan with fat, and cooks thin or small pieces over high to medium-high heat, browning them on one side before they're flipped, stirred, or shaken in the pan. Sauté vegetables and meats and you'll use very little fuel, and very little oil. Some cooks brown the exterior of thick foods (like a salmon steak) in a skillet, then "bake off" the pan in a hot oven to cook the interior all the way through. Instead of wasting oven fuel, slice the food thinner and sauté. You can also brown the food, then add a lid to create an oven effect in the pan. After sautéing, if you stir up the browned bits from the bottom of the pan with a little liquid, you'll have a tasty, natural sauce.

## Shallow Frying

This is the happy medium between deep-fat frying and sautéing. Shallow frying creates crisp exteriors like deep-fat frying, but with less heat and less oil to dispose of later. Even a small amount of oil helps conduct heat and crisp the outside. Foods cook fairly quickly and need to be flipped midway. Shallow-fry foods that are one inch thick or less, like pork chops, boneless chicken breasts, and breaded eggplant slices. Go lean on the oil; when you add the food, the oil will bubble up higher than you might expect. Oil filled to a third or a half of the depth of the food is all you need.

## Deep-fat Frying

It's hard to resist crispy tempura, onion rings, or the crunch of Southern fried chicken (though your cardiologist may disagree). If it wasn't for that pesky issue of where cooking oil comes from (how it's grown, packaged, and transported) and how to dispose of it, deep-fat frying would have a fairly green cookprint: it's fast, fuel-efficient, and can cook a wide range of foods, though for most folks, it's something to eat in moderation.

## Greener Ways to Fry

- Drain on old cardboard or paper: instead of draining fried foods on layers of clean paper towels, use absorbent paper products you might normally throw away. Cut panels from cereal boxes, used printer paper, magazines, or old newspaper (a British favorite for fish and chips). If you're concerned about inks touching your food, a single paper towel can act as a barrier.

- "Small" fry: smaller and thinner pieces cook more quickly, whether you're deep frying, shallow frying, or stir frying.

- Vent the air: a kitchen exhaust fan sucks up grease and fumes, so fewer airborne particles settle on kitchen surfaces. This means less greasy dust and less need for cleaning over time. But don't run the fan longer than needed, to conserve power (for more on kitchen fans, see page 32).

- Use a splatter guard: this low-tech mesh-screen gadget catches larger particles of grease and helps prevent the smaller ones from circulating wildly throughout your kitchen. The finer the mesh, the more particles will be trapped. The most practical ones allow you to watch the food as it cooks, so you don't have to lift them to peek; silicone models tend to block your view.

- Pick the right oil: opt for oils with high smoking points, made from sustainable crops that enhance the global good, rather than expensive biofuel crops.

- Recycle your oil: find a local program to recycle your oil into biofuel. If this is not possible, dispose of it properly: seal in a nonrecyable container and send it out in the trash. Don't pour it down a drain.

- Consider an electric fryer: if you use it often enough to offset its footprint, an electric deep-fat fryer regulates the energy it uses to keep the oil at the ideal temperature, has a filtered lid to prevent fumes and splatters, emits less heat into the room, and heats the oil more efficiently than a pot on a burner.

Deep frying submerges foods in extremely hot fat (usually 350 to 375 degrees F). Instantly, scalding heat transfers to every bit of the food's surface and rapidly penetrates the interior. In just a few minutes, small foods like shrimp are done; a bucket's worth of fried chicken pieces take just twenty minutes. Even when heated to more than 100 degrees above the boiling point of water, oil hits its mark faster than water will boil.

*Tip:* When the oil's cooled, you may be stuck with enough quarts to run Willie Nelson's bus around the block a time or two. This is actually a good way to use your old frying oil: recycle it into biofuel. Modified diesel car engines can go as much as 50 miles on a gallon of vegetable oil (fresh or used). Drivers of these cars will eat your oil, gladly. Or, ask a tempura or other fry-happy restaurant if it recycles its oil; you may be added to its regular oil recycling pick-up. Community recycling programs sometimes repurpose it. Whatever you do, don't pour frying oil down the drain; it's not good for the planet. At the very least, pour it into a nonrecyclable container and throw it out with the trash.

## Smoking Points and Refrying Your Oil

Is it okay to fry in the same oil more than once? And how hot can oil get before it starts to smoke? The answers to both these questions depend on the type of oil. Chemically, heat causes all oil to break down, lowering its smoking point. Some oils break down with just one heating, and others can last a few rounds. The optimum oil temperature for deep frying occurs between 350 and 375 degrees F. To get the most mileage out of your frying oil, use oil with a high smoking point and avoid letting the heat ever reach that point. (Overheated oils don't just change in flavor; there's some evidence that superheated oils create free radicals, which may lead to carcinogens.)

To save oil, let it cool, strain it through a fine sieve or coffee filter, and store it in a cool, dark place; if after the first round of frying it starts to smell unappetizing, stop using it. Peanut oil is one of the best frying oils because it has a high smoking point (450 degrees F). Olive oil smoke points range wildly from 220 to 460 degrees F. depending on the level of processing; cold pressed extra-virgin olive oil typically has a low smoking point, making it unsuitable for high-heat frying, while "light" olive oil can reach 450 degrees or higher without smoking (the label may specify the smoking point). This

## Dry-Fry Flourishes for a Toasty Pantry

If you're going to spend time and fuel getting a pan hot, make the most of it before jumping into the recipe. Cast-iron and other heavy pans can take a while to heat up, but once hot, retain their heat with less fuel than flimsy thin pans. You can dry-fry, or toast in butter or oil, handy ingredients for later use. Simply wipe the pan dry (careful: it's hot), then proceed with the recipe you first planned to make. For instance, get a jump on an Indian spice mix, or toast enough nuts to last a season. Toasted bread crumbs sprinkled on vegetables elevate texture and flavor. What can you dry-fry or toast in a hot skillet?

- Nuts: toast dry or in a small amount of oil, then wipe the pan dry; freeze up to six months to toss into, or onto, everything from soup to dessert.

- Seeds: sesame for Asian cooking and breads; sunflower and pumpkin seeds for salads, breads, or snacking.

- Spices: cumin, fennel, cardamom, and caraway seeds.

- Croutons and bread crumbs: cook dry or in a small amount of oil or butter and wipe the pan dry after toasting. Toast panko bread crumbs (large flaky crumbs) for toothsome crunch.

- Garlic cloves, in or out of their peels: toast until golden; peel before use.

- Grains: toast barley, quinoa, or rice, and store tightly sealed until ready to use (within two weeks for best results).

If not using them right away, store your toasted goodies in recycled glass jars in the fridge, or freeze them for longer life.

isn't a complete list, but here are a few reliable frying oils to consider, with their smoking points:

Peanut, soybean, and safflower oils: 450 degrees

Canola, grapeseed oils: 435 degrees

Corn: 400 degrees

Sunflower: 390 degrees

New oils are hitting the shelves, including refined avocado oil with a smoke point of 520 degrees (and a price point to match), followed closely by tea seed oil at 486 degrees. You can also find oil blends (like canola and corn, for instance). Corn oil used to be affordable, but with the demand for corn-based ethanol, corn oil prices are on the move. As the landscape for biofuels changes, ecovores may need to research which oils are more economical and environmentally preferred. When important food crops become fuel crops, the benefits are cloudy at best.

## A Little Oil Speeds Heat

Because fats conduct heat so well, adding a thin layer of fat to foods helps them cook quicker. Lightly oil peppers before roasting them, and the skin will blacken more evenly and faster. Roasted vegetables and seared meats profit from a light veil of oil, which hastens browning and improves flavor. A pan with a hot, thin film of oil in it cooks foods faster than a dry skillet.

# SIX

# Cookware and
# Fuel Efficiency

*In Suzy Q's wedding registry, she asks for a skillet that will last a lifetime
(that's how long she expects her marriage to last). Cousin Carla gives her a
fancy "Uber-Clad" pan, with three layers of aluminum sandwiched between
stainless steel. Aunt Bee gives her the same pan, but hers is lined with Teflon.
Uncle Buck proudly picks his favorite campfire pan, a hefty cast-iron skillet,
and Little Nicky ties a bow on a bright, shiny, easy-to-lift all-aluminum frying
pan. Based on these skillets, how long does each relative expect Suzy's marriage to last: two years, five years, possibly a lifetime, definitely a lifetime?*

    *Answer: Carla's clad pan with a stainless-steel exterior could possibly last
a lifetime. Aunt Bee's Teflon coating will bubble and blister and show signs of
wear within two years, limiting the usefulness of the entire pan well before
the rest of the pan's demise. Nicky's all-aluminum pan will show pitting from
salt and acids within five years. Uncle Buck's cast-iron skillet will, with
proper maintenance, last not just for a lifetime of marriage but can be passed
on to Suzy's children and grandchildren when she dies. To find out what
makes the difference in these pans, read on.*

## Cookware Considerations

Good cookware should last forever. Every time you toss out an old pot or
pan, and bring in a new replacement, you're fattening your cookprint. But
durability isn't the only important attribute; you also want to consider the
cookware's energy efficiency, broadening your search to include cookware
that may be old-fashioned or even "pre-owned." With so many surfaces,

materials, and price ranges, how do you know which pots, pans, and oven-ware will perform best over time?

If you've already got a good set of pots, pans, and utensils, then buying a new set just enlarges your cookprint and dumps more "stuff" into the world. But if your cookware is inefficient, warped, or badly worn, follow this chapter's tips when selecting new pieces to add to your toolset. (And donate, recycle, or repurpose your old cookware; one man's trash is another man's treasure, and even funky old pots can be reborn as animal feeders, bird baths, and planters.)

Consider, too, how and where the cookware is made. Does the factory follow green practices? Does it pollute the air and waters of a foreign country before being shipped to our shores? Well-designed cookware from overseas may not have a domestic alternative. Choose wisely, for cookware that lasts a lifetime, or close to it, may cast a relatively small cookprint despite a transcontinental voyage. New brands with greener materials are surfacing daily, so be prepared to dig a little for the best choices. (NewGreenBasics .com posts frequent cookware news and reviews.)

## Green Cookware = Fuel Efficiency

You've heard the saying "pick the right tool for the job." When you match the right cookware to the job at hand, you create a more energy-efficient cookprint. But this means asking a few questions. Are you cooking low and slow, fast and hot, or somewhere in between? Are you cooking on top of the stove or in it? What exactly does "clad" mean? Is a baking dish better if it's metal or glass?

A cookware's performance starts with its material composition. Without getting too technical, you should evaluate what you need based on these four properties:

- Thermal conductivity: a material's ability to conduct heat (that is, the speed with which it moves the heat around); different materials conduct heat faster or slower than others.

- Heat capacity: a material's ability to store heat (technically, the amount of heat required to raise the temperature of a substance by one degree Celsius); some materials store, or hold onto, heat longer than others.

- Responsiveness: the material's ability to respond to changes in the heat source.

- Reactivity: a material's susceptibility to chemical changes; reactive cookware typically responds to materials like acids and salts. Many reactive materials are coated to prevent chemical changes from occurring (making them chemically inert or inactive).

If this muddles the brain, relax—the rest of this chapter converts thermal theory to cook-pan reality, looking in turn at different kinds of skillets, woks, Dutch ovens, haybox ovens, pressure cookers, and bakeware.

## Picking a Skillet

Here's the perfect skillet for your kitchen: it's got a thick, heavy base that conducts heat well. It distributes heat evenly across the entire pan surface and up the sides of the pan (so it sears and browns foods beautifully). It responds readily to adjustments in burner settings, and it makes productive use of almost all of the heat generated by the burner. Salt or acids don't affect the material, and the surface withstands metal and sharp utensils. The pan comes with a clear lid, so the cook can watch the food cook. It's truly nonstick (foods slide right off), easy to clean and maintain, nontoxic, and safe at very high heat. The pan and its handle are ovenproof to 600+ degrees F. It works on conventional and induction burners. For weak-wristed cooks like me, it's also lightweight enough to use daily, without strain.

Such a pan does not exist.

Many traditional skillets come close (new cookware lines are coming even closer). But like humans, no pan is perfect—either in its cooking abilities or in its green cookprint. Your best bet is assembling the perfect *combination* of skillets. An efficient cook really needs more than one type of skillet, in large and small sizes. Generally, a 12-inch and an 8-inch skillet will get us through most recipes.

If you invest in quality, rather than repeatedly buying and tossing away poorly made skillets, you'll consume less. And a well-made, efficient skillet helps conserve fuel. Even if a good skillet doesn't last a lifetime, ten years or more is a reasonable warranty. Don't expect a cheap skillet to perform well, or last long. When you're playing with fire, you need durable materials, not disposable ones.

In short, the ideal skillet conducts heat well, responds rapidly when you adjust the burner, heats up evenly with no hot spots, and can handle very high temperatures. Some cooks would add nonstick or easy release qualities, and easy-clean surface. You'd have to buy two or three different types of skillets to meet all of these criteria: (1) a sandwiched metal-clad skillet, (2) a cast-iron skillet, and (3) a nonstick skillet.

As a personal choice, I enthusiastically recommend the first two skillets, but I have reservations about the third. The metal-clad skillet handles high heat well, and the cast-iron skillet is even better, reaching super-hot temperatures, though both come with drawbacks. A Teflon-style nonstick skillet is convenient, but toxicity issues muddy its cookprint; ceramic-based nonstick pans make great, safe alternatives.

## Metal-Clad Skillets

Cooks need a skillet that distributes heat evenly, because the average burner is anything but evenly heated. Electric stoves have coils, and gas burners have a ring of flame. To make up for the hot and cool gaps in the burner's design, you want a skillet base that absorbs and distributes the heat throughout the metal bottom, not just where the heat source meets the metal. Copper and aluminum are very good heat conductors. But copper is expensive, and aluminum is soft and reacts with food acids. Clad pans make the most of multiple metals. The best ones sandwich one or more layers of copper or aluminum between a tough outer surface, like "18/10" stainless steel (18 percent chromium, 10 percent nickel), or hard anodized aluminum. Cheaper skillets sandwich layers so thin you don't get much benefit. So pick a skillet with a hefty bottom, at least 1/8th of an inch thick. Other good features: handles that can go from cooktop to oven and withstand temperatures of 500 degrees or higher, a clear lid, and a lifetime warranty.

## Cast-Iron and Cast-Iron Enameled Skillets

Cast iron takes a long time to heat up, but once hot it retains the heat well and distributes it evenly. It also takes a long time to cool down, so you can turn off the heat early and let foods passively cook in it. With proper care, the surface becomes increasingly stick-resistant over time. Cast-iron pans with baked-on enamel are even more stick-resistant, and unlike traditional cast iron, the surface needs no extra care or "seasoning." Cast-iron cookware comes in all shapes and sizes, from deep Dutch ovens to tiny skillets and griddles. But there are a few drawbacks: cast-iron pans are heavy to lift, they can scratch glass cooktops, and enamel surfaces can chip. Enameled pans don't react with acids, but iron surfaces do react to tomato sauce and other acids, which may affect the flavor and color, though it's not harmful. On the plus side, these pans will literally last generations with proper care, and many chefs swear by them as a preferred type of cookware. They withstand very hot heat in the oven and on the cooktop. They're also very affordable, though enameled pans do cost more. Look for USA-made cookware. (Lodge brand makes its cast iron in the USA, but the company's enameled pans come from China.)

### To Recirculate an Iron Pan

Even a poorly cared for, rusted cast-iron pan can be restored easily by scraping and washing the rust away, oiling the surface, and baking it in the oven or barbecue. Shop for used pans in thrift stores and put them back into service, for a fraction of their original or replacement cost.

## Nonstick and Ceramic-Lined Skillets

The pan itself should have a heavy bottom and be made of metals that conduct heat well, but the real issue is whether the nonstick surface is safe. Nonstick coatings, which come under many brands besides Teflon, are applied to all types of clad and nonclad pans. At high temperatures, the surface starts to degrade and may cause health issues. Because Teflon-style pans are so widespread and appealing (you probably own at least one), they're addressed in detail in the next section, but first, consider a greener alternative.

Hard anodized aluminum is electronically treated to be extremely hard and durable, with the good conductive quality of regular aluminum. It's not nonstick, but it resists sticking, so it's a popular cookware material. Now, it's even better. Anodized clad pans with an aluminum alloy core and a ceramic-based surface that is both nonstick and nontoxic have recently come onto the market. The cookware (Cuisinart's GreenGourmet is a leading brand) contains no petroleum products, no PTFE or PFOA, can withstand oven use up to 500 degrees F, and is broiler-safe. Some pans conduct heat so well that

they perform best when *not* used on high heat; medium and low are sufficient (meaning built-in energy efficiency). Drawbacks: they're typically
made in China, yielding a traveling cookprint similar to most cookware
these days, and they don't work on induction burners (no ferrous material).
Unlike untreated anodized aluminum, the ceramic-based surface can chip,
but so can enameled cast iron (which is like lifting an anvil, by comparison).
Avoid sharp or metal utensils (anything safe for Teflon is fine). This cookware should last with proper care, but it's too soon to know if it has a lifetime profile.

### Teflon-Style Skillets: Choose to Use, or Lose?

*Beware of any manufacturer's claim that the FDA has "approved" or has
certified a coating. Nonstick coatings can be comprised of ingredients
that are "generally recognized as safe" (known as the GRAS list) but the
FDA does not test, certify or otherwise approve any coatings applied to
noncommercial housewares products.*

*—Cookware Manufacturers Association*

*Good Housekeeping's experts say you can use nonstick safely, as long as
you use it properly. Any food that cooks quickly on low or medium heat
and coats most of the pan's surface, bringing down the pan's temperature, is unlikely to cause problems.*

*—Environmental Working Group*

Some experts say nonstick pans are safe "when used properly," but details I
discovered while researching this book remind me of the film *Thank You for
Smoking*. Are these pans safe or not?

Nonstick surfaces contain a chemical polymer known as PTFE and come
under many variations and brand names, with Teflon being the most well
known. The Environmental Protection Agency and DuPont (which makes
Teflon) have long downplayed any risks associated with nonstick cookware.
But the manufacturing process creates enough toxins to sicken workers, and
PFOA (perfluorooctanoic acid, a key ingredient) has been shown to harm
birds and small animals, is linked to low birth weight in humans, and is per-
sistent in the environment. PFOA is being phased out by 2015.

The convenience of nonstick cooking is hard to pass up. For a while, I
simply rethought the ways I used nonstick pans. I pulled them out only

when the nonstick feature made a real difference, and I vowed to cook gently on medium or low heat (never on high heat).

Nonstick coatings are said to be safe if used according to the manufacturer's instructions, which include not heating an empty pan, and never cooking over high heat. (PTFE begins to degrade at 500 degrees F). But let's face it: most cooks love to fry over high heat (especially if the recipe says to), and if you want to cook quickly or stir-fry, then high heat is a must. Without it, moisture is coaxed out of food and the surface never browns properly. (By the way: PTFE isn't just for cookware. It's in virtually every industry. It's the magic behind high-performance fabrics like Gore-Tex outdoor wear, Space Shuttle parts, machinery lubricants, and gaskets for trains, planes, and automobiles.)

Besides creating fumes, high heat causes nonstick surfaces to blister and flake. Once the surface starts to peel and bubble, it's time for a new skillet. Even with proper use, the best-made nonstick surfaces start to degrade within two to five years, and with cheaper nonstick skillets, the surface starts to blister the first few months. Manufacturers say the peeled coating is inert and will pass through your system without risk.

After decades of disposing of deteriorating nonstick skillets, I've stopped using them entirely. What we do know is enough to scare me, and what we may not know about the long-term environmental effects concerns me even more. So now I reach only for ceramic-lined, metal-clad, or cast-iron pans without conventional nonstick coatings.

If you own nonstick cookware and like using it, then Good Housekeeping outlines these precautions:

- Never preheat an empty pan.

- Don't cook on high heat.

- Ventilate your kitchen.

- Don't broil or sear meats.

- Choose a heavier nonstick pan.

- Avoid chipping or damaging the pan.

Personally, I've decided that for my family and the planet, nonstick coatings simply aren't worth the risk. This is cookware we can live without.

Fortunately, other methods and safer cookware still offer options for cooking in a nonstick style.

### How to Cook Without a Nonstick Skillet

To prevent food from sticking to the pan, the Chinese have a saying: hot wok, cold oil. It refers to carbon-steel woks, the traditional kind without special coatings, but it applies to all metal cookware surfaces, even barbecue grills.

Follow this four-step process to keep food from sticking:

## Pick a Pan, Not Just Any Pan

What you're cooking determines which type of pan to use for the job. As discussed earlier, metals may heat up quickly and cool down quickly, or they may heat up slowly and hold the heat for a long time. When buying and using stovetop cookware, consider these other points:

- Size: match the pan footprint to the burner size. Too large or small a pan wastes fuel.

- Inside size: skillets are measured across the top diameter, but the interior bottom diameter is where the food cooks. An extra quarter-inch or half-inch makes a difference (chicken pieces, for instance, will steam instead of brown if too crowded).

- Flat bottom: warped bottoms absorb heat unevenly, wasting fuel; thin pans are more apt to warp.

- Thick bottom: for even heat distribution, choose thick over thin (thin bottoms allow for hot spots).

- High heat delivery: to brown foods efficiently, you need a pan that conducts high heat to the food. Very efficient skillets may need only a medium heat source to produce a hot surface. Others require the burner be set on high to reach a high cooking

*(continues)*

1. Heat the pan. When water drops dance across the surface, it's hot; after the water evaporates,

2. Add at least a thin film of oil and swish it around to coat the surface; with grills, oil the surface of the food.

3. Add the food.

4. Wait; don't move the food for at least ninety seconds, or until lightly browned, at which point it will release tension from the metal and be able to be moved.

---

(Pick a Pan, Not Just Any Pan, *continued*)

temperature. The better the heat capacity, the longer the pan will hold onto its heat, and the less fuel used during cooking.

- Safety at high heat: Teflon-style nonstick cookware is not safe at high temperature. But new forms of nonstick surfaces (like ceramic-based linings) are safe even over high heat.

- Nonreactive or reactive: some metals react to acids, meaning tomato sauce will turn brown or develop an off-flavor. Reactive pans include cast iron and aluminum, which are good conductive metals. The reactive effect won't harm you, but if the recipe calls for a nonreactive pan (or uses a lot of acidic ingredients), then pick a nonreactive material like ceramic-lined cast iron.

- Ovenproof pan and handle: to be able to brown foods over a burner and transfer the pan to an oven (or broiler) to finish cooking, you need an ovenproof pan and handle. Metal handles get hot but withstand very high temperatures; nonmetal handles may be safe up to 400 or 500 degrees F (check the label), and others are not ovenproof.

- Dishwasher safe: not a deal breaker, but if washing pots and pans in a dishwasher is important (saving on water), then check the label. Some cookware deteriorates or discolors when machine washed.

Enamel-coated pans are more stick-resistant than metal surfaces, but even they perform better if you follow the same procedure.

## The Indispensable Wok

*A wok, a round-bottom and now also available flat-bottom, bowl-shaped pan, has been used for centuries in Chinese kitchens. This pan was designed out of a necessity to conserve on fuel because a wok conducts heat very efficiently [and] uses up little kitchen space since a wok takes the place of many pots and pans. With your wok you can stir-fry, steam, braise, sauté, simmer, deep fat-fry, stew and even smoke. The advantage of using a wok rather than a skillet is that the flared sides and depth permit rapid tossing of many ingredients and the use of a minimum of oil. The bottom of the wok gets extremely hot and the sides remain cool, thus, the food is cooked in the bottom and tossed to the sides; with this constant motion, vegetables are cooked just right . . . in the hot bottom, then tossed to the cool sides—quickly, efficiently, nutritiously!*

*—The Wok Shop in San Francisco*

Woks are models of energy efficiency: they require high heat but cook quickly, and their design maximizes the heat's distribution. The classic Chinese wok is made of carbon steel and, like cast iron, requires proper care and "seasoning" to develop a patina that protects the surface and develops "wok

### Go Faster! Three Speedy Cooking Tips

- Pound, butterfly, or slice meat to increase the surface area and shorten overall cooking time.

- Reduce a liquid for a sauce by using a large skillet rather than a pot. By exposing a larger surface area, the flavors concentrate more quickly with less fuel.

- Roast meat and chicken pieces on a rack or grid in a low-sided pan, like a sheet pan. Heat penetrates from all sides, so the food cooks quicker and more evenly.

hee," the breath of the wok. The wok's design is elegant, in the scientific sense of the word, meant to rest on a collar over a hot flame or gas burner, and some designs are modified to work on flat cooktops. Carbon steel outperforms newer materials but needs some maintenance. For maintenance-free cookware, consider a five-ply wok with stainless exterior and three interior layers of aluminum and/or copper. Enamel-coated iron woks with flat bottoms perform well on electric cooktops, which need help in delivering adequate heat (iron takes longer to heat but retains heat well once hot). Stay away from nonstick woks: their safe usage can't be trusted for high-heat cooking.

## The Double-Duty Dutch Oven

In the arsenal of fuel-efficient cookware, the Dutch oven is essential. The Dutch oven is a heavy, deep pot with a lid, designed to retain heat and distribute it evenly. It evolved from cooking in hearth fires; the lid had a rim so hot coals could be placed on top of the pot, essentially making the vessel into its own oven. (The first Dutch ovens also had looped handles, so they could be hung from a chain or a rod over a fire.) Dutch ovens can be used in the oven or on a cooktop, making them double-duty functional. Enamel-coated cast-iron ones are my favorites, as they don't react with tomato sauces and vinegary-braises and are easy to clean. They come in sizes small enough for a night of rice or large enough to cook a goose, and in round, oval, and even square shapes. For me, the most versatile pieces are a small round vessel (around 3-quart size) and a large oval casserole (around 7-quart size), both enamel-coated and with ovenproof handles and lids, so they can go from cooktop to oven. As the next section on haybox cooking explains, a cast-iron Dutch oven's insulating properties can stretch fuel even further.

## Retained Heat Cooking: Haybox Ovens and Modern Thermal Ware

*We cook casseroles and stews in an old bedside cabinet we have converted to a "haybox" cooker. We filled it with offcuts of Celotex insulation trimmed to tightly fit a Le Creuset casserole (it has to be cast iron). We bring the casserole to the boil on the stove (allow 20 minutes on the stove for meat dishes), slide it into the "cooker," fit in two extra blocks of insulation around the pot to completely surround it, and close the door. The residual heat cooks rice in 30 minutes and meat casseroles in three hours.*

—George Marshall, of the Yellow House project
in England (www.theyellowhouse.org.uk)

The concept is elegantly simple: partially cook foods in a tightly covered vessel, put the hot vessel in a well-insulated space, and let the retained heat continue cooking the food. The process consumes up to 80 percent less energy and can be used with beans, grains, stews, and roasts.

Whenever fuel is scarce, people work hard to conserve it, which often leads to very logical inventions. Haybox cookers have stretched fuel in World War II England (hailed as the Victory Oven), sparse African villages, and American pioneer settlements and on Boy Scout camping trips. Cast iron is the favorite cookware material, but in practice, any good heat-retaining cookware can be used. Cultures without iron rely on earthenware.

Today, haybox cooking is being revived by green-conscious cooks and off-the-grid gurus. I've seen DIY versions of haybox cookers made with ice

## Ovenproof Handles and Lids

Invest in cookware that can be used both on the stovetop and in the oven, to reduce the number of pieces in your arsenal. Many otherwise terrific skillets have handles that can't take an oven's heat, limiting their versatility. Cast-iron skillets can go from cooktop to oven (but not the microwave), so they're good for this type of cooking, and metal-clad skillets with heat-resistant handles (to at least 500 degrees) work well, too.

Speaking of handles, I'm leaning increasingly to pans with two short handles on opposite sides, rather than a single long handle. Short handles, or assists, are easier to store, take up less space, and are safer, too: long handles are more likely to be bumped or knocked off the stove. The more compact the items, the less cabinetry needed in a kitchen to store all that kitchen stuff.

chests and blankets or holes dug in the ground, and this book's blue oven strategy (where the oven's heat is turned off before cooking is done) is yet another variation of retained heat cooking. (www.LostValley.org has a detailed plan for making your own haybox cooker; one guy simply uses towels and a thermal survival blanket.)

The hip, modern version of a haybox is Durotherm and Hotpan thermal cookware (made by the Swiss company Kuhn Rikon): high-end stainless-steel pots with insulated convex lids (which self-baste the food) and an insulating shell that "cooks" by retaining heat, without added fuel. The food never burns, remains hot for up to two hours, and uses up to 70 percent less energy than traditional methods. Simmer brown rice for 10 minutes, then transfer the pot to the thermal shell for 30 minutes. Simmer polenta for 1 minute, then let it passively cook for 20 minutes. Sizes range from 1 to 3 quarts. The Hotpan's outer shell (in assorted colors) isn't a one-trick-pony either: it works as a salad or serving bowl and keeps breads warm or ice cream cold.

## Pressure Cookers: All-Purpose, All-Season

Pressure cookers and pressure cooker books seem to be promoted most heavily in winter. I suppose this makes sense, given that pressure cookers whip up beans, stews, soups, and other steamy hot meals with enormous ease, in a fraction of the time spent by conventional cooking.

My pressure cooker comes out most frequently in summer, when I'm struggling to keep the house cool. A pressure cooker gets me in and out of the kitchen fast, without much ambient heat. Meaty stews aren't on the summer menu, but grain salads, herbed risottos, fresh vegetables, and quick meals are.

The pressure cooker works like a gem with summer's fresh bounty, especially the bushels of tomatoes, zucchini, eggplant, and other organic produce exploding out of the garden. You save some time with fresh veggies, but they also retain more nutrients and flavor when pressure-cooked. And you really save time and fuel with grains, lentils, and beans—which make fabulous salads when mixed with fresh herbs and vegetables and are a welcome change from the pedestrian pasta salads of summer. Of course, the pressure cooker is super-energy-efficient for wintry braises and stews, too, so pressure cooking is one of the best green strategies all year round. (They even make cheesecake; no oven required.)

## How a Pressure Cooker Works

Pressure cookers are like steamers on steroids. Lock the air-tight lid in place, set the pot over high heat, and allow the liquid inside to convert to steam. As more steam is created, the internal pressure builds (to 15 pounds per square inch, or PSI; an indicator on the lid lets you know when high pressure is reached), and consequently, water's boiling point jumps from the normal 212 degrees F at sea level to 250 degrees. Within this hotter-than-normal temperature, food cells break down and cook in about one-third the standard cooking time. Depending on the recipe, you bring down the pressure manually by running cold water over the lid, or by using a special release lever on some cookers; or do nothing and let the pressure gently subside on its own, which takes a few minutes. Pressure cooker recipes tell you which release method to use.

Best candidates for pressure cooking: soups, broths, stews, braises, chili, pot roasts, brisket, corned beef, lamb shanks, chickens, potatoes, winter squash, whole grains, rice, risotto, one-pot meals (like paella, jambalaya, and biryani), beans, beets, pasta sauces, bread pudding, fruit compote, and even sweet surprises like no-oven cheesecakes, pudding cakes, and custards. For recipes and more tips, type in "pressure cooker recipes" at GlobalGourmet.com.

## Before You Buy a Pressure Cooker

Horror stories of exploding pressure cookers belong in the past. Modern units build-in multiple safety features (including a three-stage back-up system of regulators, vents, and expanding gaskets), so don't let fear of pressure cooking prevent you from tapping into the freedom these ingenious devices deliver. Today's pressure cookers are 100 percent safe.

Use pressure cookers as regular pots and pans, too. Just because they have locking lids doesn't mean you have to cook under pressure. (Scale back on your cookware inventory this way.) You can brown and sauté in them, and then cover and cook under pressure if desired, or not.

Personalize your perfect size. Most families will prefer a 6-quart capacity (it's also the standard size for most recipes). An 8-quart unit is good for quantity cooking. For couples and smaller recipes, consider a 2-1/2 quart pressure fry pan, or a braiser with two assist handles. Remember when sizing that you only fill the cooker to two-thirds capacity: a smaller braiser may be more energy-efficient, since it reaches pressure more quickly, but a 6-quart unit is more versatile if you're buying only one size.

Buy for the long haul. The best brands come with warranties of ten years or more but should last far longer with good maintenance and a new rubber gasket for the lid every now and then.

*Buying tips:* European brands measure in liters, about the same size as a quart, and make some of the highest-quality pressure cookers. The more established the brand, the better chances of finding parts in the next decade or two. Opt for quality over price: performance matters. Make sure the model cooks at a full 15 PSI (some don't, and thus offer no real energy savings). The pot must have a heavy bottom for even heating (stainless-steel clad, with an aluminum or copper core). I've been very happy with Kuhn Rikon and Fissler, but consider all brands and models to find the best pressure cooker for you.

## Ceramic Bakeware

If you've ever walked through an ancient civilizations exhibit at the Metropolitan Museum of Art, you know that pottery cooking vessels can, if not shattered into dust, survive thousands of years.

Today's family of modern pottery cookware also lasts long, and if a piece breaks, no worries. Ceramics (the family of glass, clay, porcelain, stoneware, and earthenware) are fairly environmentally friendly and have a smaller manufacturing footprint than other cookware. Ovenproof glass cookware is made from sand. Porcelain, stoneware, and earthenware are crafted from types of clay, and include Römertopf bakers, pizza stones, Moroccan tagines, bean pots, Chinese sandpots, and even terra-cotta flowerpots.

They're true earth materials, though some additives are mixed in or glazed on for color or thermal properties (some imports may contain toxic glazes, so check the label to make sure they're food-safe). Though durable, ceramics can break or chip, so they may not last as long as metal, but then again, they also break down more easily in the environment. Dust to dust, as they say.

Ceramics withstand high oven temperatures but aren't meant for cooking over direct heat. They're made by countless brands. Pyrex and Anchor Hocking are the most common names of glassware in the United States, and CorningWare is famous for its white stoneware casseroles, ramekins, and baking dishes.

## Special Attributes of Ceramics

Ceramics offer bona fide benefits when it comes to oven cooking. They're poor heat conductors and good insulators. This sounds underwhelming, yet it actually forms a winning combination for casseroles and baked goods.

Ceramic vessels absorb heat slowly, but they generously share the heat with the food, letting it slowly penetrate into a casserole (rather than browning on the edges before the center is cooked, which can happen with metal baking dishes). Once hot and out of the oven, they slowly release the heat, making them good insulators for holding that casserole at a warm and cozy temperature. They're ideal for blue oven methods of starting in a cold oven, or finishing in a hot oven with the heat turned off.

Other reasons to love ceramics:

- Glass baking dishes are clear. You can watch the sides as juices bubble up or crusts brown, so there's no need to open the oven door and release heat just to peek on progress.

- Smaller pans are versatile and cook foods faster than larger pans. You can roast or bake almost anything in a glass or ceramic pie pan or baking dish. (I roast chickens in pie pans.) Pie pans and 8-inch-square pans slide easily into toaster ovens and microwaves, as do smaller rectangular dishes. *Tip:* Instead of cooking one large casserole, divide the ingredients into two smaller pans, like deep-dish pie pans or 8-inch-square baking dishes, and shorten cooking time by 5–10 minutes. If you don't finish both pans that night, the leftovers will be easier to store, and you can reheat them fully or partially in a microwave oven or a toaster oven.

- Ceramics are multifunctional. Refrigerate leftovers in them, and reheat in a conventional, toaster, or microwave oven. (Check the manufacturer's usage recommendations first.) Buy ones with microwave-safe rubber lids or glass lids, and skip covering them with plastic wrap.

- Ceramics are "low-stick" and clean up easily. They're nonreactive, nontoxic, and don't absorb odors. Gentle soaking releases baked-on food, and white vinegar removes tougher spots.

## Don't Make 'em Like They Used To

Old CorningWare is not the same as new CorningWare. Introduced in 1958, the original pieces were made of glass-ceramic and were durable enough to work on the stove, in the oven, and under the broiler. In the 1990s CorningWare's ownership changed, and so did the material. The pieces are no longer recommended for broilers, and especially not for cooktops. If you run across older pieces (especially those with the cornflower blue emblem) at garage sales or thrift stores, snatch them up. They may outperform their modern-day counterparts (and they're very hot as collectibles, too).

*Caution:* Pyrex baking dishes have been a household staple for generations. Over the past decade, some Pyrex pans have shattered coming out of or even in the oven. Interestingly, these problems popped up after World Kitchen bought the Pyrex company in the 1990s (it also bought CorningWare), suggesting that a change in manufacturing may be at root. World Kitchen says the cause is thermal shock, and recommends not placing hot pans on cooler surfaces, and adding a small amount of liquid before putting in a roast. I haven't experienced problems, but cook with care whenever you use glass baking dishes, and follow the instructions that come with the pan.

A great buy, used or new: borosilicate glass. It's what Pyrex was originally made from, and it's the glass used in chemistry labs, designed to be strong and withstand inferno-range temperatures. True borosilicate cookware isn't cheap, but it's safer, stronger, and lasts far longer. (Some European brands make true borosilicate cookware.)

## Silicone Bakeware

Silicone cookware is essentially synthetic rubber. It's made from silicon, a
natural element found in sand and rock, which is bonded to oxygen and
then dabbled with to create everything from liquid to solid substances, each
with varying properties, and considered stable and chemically inert (making
it nonreactive as a cookware). The reusable, ultra-slick Silpat sheets, made
of silicone, let baked goods slide right off, and can replace disposable parchment paper in many instances. Most other silicone cookware is stick resistant, but it's not nonstick. The Food and Drug Administration considers pure
food-grade silicone safe, yet cheaper brands add bulk fillers (pinch and twist
a flat area of the silicone; if it turns white, it's got fillers in it). Silicone utensils like spatulas won't scratch nonstick surfaces, but silicone vessels are not
meant for cooking over direct heat. Silicone cookware varies in performance, and the pieces don't respond in the same way as glass, ceramic, or
metal bakeware. Some cooks adore silicone pans; others detest them.

China produces almost all the silicone cookware currently available on
the market. Not a lot is published about the manufacturing process there,
but Chinese workers have a high rate of a fatal lung disease known as silito-
sis, caused by fine dust particles of silicon, granite, and other mined or
ground substances. Silicone leaves a carbon shipping trail from overseas,
though the material is so lightweight, it's less of an issue than with metal
cookware made in China.

## Metal Bakeware

As with other cookware, buy pieces that will last. If the pan feels flimsy, the
quality will be, too. The greenest cookware will always be the one that uses
fuel most efficiently relative to what you're cooking, with the tastiest results.
(You could bake cookies on cast iron, but what a waste of fuel that would
be!) Lined copper bakeware is the most fuel-efficient option, but it's priced
beyond most household budgets. Of the more affordable metals, consider
these options and their attributes, so you can pick the greenest cookware for
the job at hand:

- Aluminum: great conductivity (it heats up quickly), but it reacts with
  acids (line it with parchment paper if roasting tomatoes, for instance).

Aluminum conducts heat really well, but thin baking sheets will warp. The best and most versatile aluminum baking sheets are known as half-sheet pans or jelly roll pans (12-1/2 x 17-1/2 inches); make sure they're heavy gauge, with 1-inch sides and rolled rims for stability. (Visit a restaurant supply store for the best bargains.) Quarter-sheet pans (9-1/2 x 13 inches) are another handy size and useful for smaller ovens.

- Anodized aluminum: conducts heat quickly and evenly. It's durable and long lasting, with a hardened surface that's nonreactive. Dark anodized pans cook in about a third less time, which may or may not be desirable for particular recipes (pastry edges may brown too fast).

- Steel: stainless steel is not a good conductor of heat, so it doesn't cook foods evenly. But when used as the outer surface of a clad pan (with an aluminum core, for instance), it makes the pan durable and nonreactive, and a good all-around choice. Tinned steel (steel with a thin coating of tin) does conduct heat well, though it takes some time to heat up and darkens with use, which can affect baking times.

- Cast iron: retains heat well but is slow to heat, heavy to lift, and reactive to acids. The section on page 94 has more details, but essentially it's best for long, even cooking.

- Insulated baking sheets: two layers of metal with air in between ensure even cooking and little browning, but this pan takes longer to heat and cook. It's not an energy-efficient choice.

Finally, keep in mind that certain small appliances may outperform conventional cookware and may reduce your total inventory, so flip back to Chapter 2 for details on cookprint-shrinking appliances.

# SEVEN

# What to Buy:
# A Cook's Guide
# to Green Basics

*While we support the growing and consumption of organic food in gen-*
*eral, it behooves us to mention that the many benefits of organic agri-*
*culture are certainly offset by the excessive amount of fossil fuel*
*required to process, package, and transport organic goods worldwide....*
*Food that is allowed to ripen in the ground or on the vine, that does not*
*undergo long trips in cold storage, and that is eaten soon after harvest is*
*fresher, more flavourful, and more nutritious. These simple pleasures,*
*along with the more serious issues like the strength of our rural commu-*
*nities and the safety and security of our food supply, are all addressed*
*by eating more locally grown organic food.*

*—Our Stance on Local vs. "Big Organic,"*
*by the Organic Food Council of Manitoba*

Do you realize we've raised an entire generation of eaters who don't understand that foods have seasons?

We're so spoiled. We enjoy juicy grapes in January, strawberries in September, asparagus at Thanksgiving, and great wines from every continent all
year round—with flavors and prices attractive enough to tempt even the
most dedicated ecovore.

The average supermarket carries more than 50,000 items. Some, like
meats and prepared frozen foods, arrive with a chunky carbon footprint; it
takes a lot of resources to create them, and even more resources to get
them into your shopping cart. Others, like locally grown carrots, use far
fewer resources, and you can eat them as-is: rinsed and raw. Large-scale

"Big Organics" producers make organic food more affordable (a good thing), but produce coming from far away may bring with it wide carbon transportation trails, competition to local farmers, diminished nutrients, and loss of freshness.

This chapter adds an ecovore's perspective to familiar green shopping guides. It conjures up questions to consider beyond what's on the label, and points out where labels fall short. New laws and standards are surfacing all the time, so even the most dedicated green cook may find a few updated issues to consider now, and to watch out for in the future.

## Don't Pick the Season, Let the Season Pick You

When your food travels more than you do, it's time to change your lifestyle. According to the Natural Resources Defense Council, whenever foods are trucked, shipped, or flown in, the environment (and your health) feels the hit, especially when they come from overseas:

> Take grapes, for example. Every year, nearly 270 million pounds of grapes arrive in California, most of them shipped from Chile to the Port of Los Angeles. Their 5,900 mile journey in cargo ships and trucks releases 7,000 tons of global warming pollution each year, and enough air pollution to cause dozens of asthma attacks and hundreds of missed school days in California. . . . The transportation-related pollution from importing Chinese garlic includes 39 times more particulate matter and 6 times more global warming impact than transporting garlic grown in California.

Compared to California produce, Australian oranges brought to the United States yield 44 times more particulate matter and 6 times more global warming impact; Thai rice has 22 times more particulates and 3 times more impact; tomatoes from the Netherlands bring 2 times more particulates and 500 times more impact; and wine from France pours out 29 times more particulate matter and 5 times more global warming impact.

Produce isn't the only foodstuff making the rounds. According to the *New York Times*, weird things like waffles and bottled water cross oceans, from one industrialized nation to another, churning up a vortex of carbon emissions in their transport. Without fuel taxes on shippers, these goods arrive at irresistibly cheap prices, even more affordable than local items, though efforts are being made in Europe to change this. Few people realize, too, that some organic foods sold everywhere from Wal-Mart to Whole Foods Markets are born and raised in China, where enforceable standards are questionable, water and air quality can be hazardous, and the distance is literally halfway across the globe. Without country of origin labeling, it's hard to tell where your food comes from, even if it's sold as organic.

The bottom line: local organic is still the best option, but it's not always within reach. If organic foods qualify for frequent flyer miles, you are probably better off buying local, whether they're conventionally grown, organic, or somewhere in between. For the concerned ecovore, the choices stir up questions like, How far has the organic food traveled (is it from California or Australia, or can you even tell)? And, If it's local and conventional, do the suppliers at least practice low-impact farming?

# The Ecovore's Checklist

Renewable, sustainable, local, organic, seasonal, fair trade—trying to find foods with all of these attributes can spark a serious bout of ecoanxiety. Relax and breathe: Even if the products you buy aren't 100 percent pristine, you can still make a difference. When deciding what food to buy, consider these aspects:

√ **Is It Organic?** Get out your gray scale for this term. The USDA defines four levels of this category based on percentage of organic content:

- **100 percent organic,** with the **USDA organic seal,** means all the ingredients are grown organically and minimally processed; they're produced without conventional pesticides, artificial fertilizers, human waste, sewer sludge, ionizing radiation, genetic engineering, or food additives. Livestock are raised without growth hormones or routine use of antibiotics. (In shorthand: no pesticides, synthetic hormones, or antibiotics.)

- **USDA organic seal** *without* the words *100 percent organic* means at least 95 percent of the content by weight is organic (excluding water and salt), with the remaining 5 percent of ingredients on an approved list of nonagricultural products not available in organic form. Products are certified (usually by a third party) and may not use excluded methods, like sewage sludge or ionizing radiation.

- **Made with organic ingredients** (without the USDA organic seal) applies to processed products that contain at least 70 percent organic ingredients, and may list up to three of the organic ingredients in the main display. These products may not use excluded methods, like sewage sludge or ionizing radiation.

- **Organics identified in ingredients list only:** processed products with less than 70 percent organic ingredients can't use the term *organic* (or the seal) on the principal label; they can only identify specific ingredients that are organic in the ingredients statement on the information panel.

- *A fifth consideration: certified organic vs. practical organic*—In the United States, the right to use the word *organic* is regulated, and smaller farmers that sell organically grown foods may not always pay the fee for the right to claim "certified organic." Instead, many market their produce as "grown without pesticides," especially at farmers markets. When in doubt, ask; most will tell you to what degree they comply with organic standards, but without third-party certification, you'll just have to trust them. Some smaller farms qualify for the *Certified Naturally Grown* label, an independent nonprofit program without the paperwork and fees of the USDA, but with the same standards.

√  **Is It Free of Synthetic Hormones and Antibiotics?** Some meat, poultry, and dairy products may be, even if they're not raised organically. Verification (under USDA organic certification or by another third-party source) helps ensure these claims are valid (e.g., using no bovine growth hormone—BGH or rBST). Factory-farm conditions promote the use of antibiotics and growth hormones. Smaller farms tend to offer more space for animals, and hence less need for constant antibiotics. Farmed fish should also be antibiotic-free (more about fish later in this chapter).

√  **Is It Sustainable?** Sustainable crops, seafood, and livestock cause no harm to the ecosystem, are not in danger of being depleted, are economically viable, and, according to some definitions, are culturally

## Ecovore's Tip: Look Beyond the Labels

Sometimes livestock and produce may be sustainable and organic without being officially certified as such. Obstacles to "humane" certification can be as innocent as keeping herding dogs on a sheep ranch. Or, perhaps the application fees are too costly, especially for small farmers and ranchers. Get to know your local suppliers. You may find a good reason why they're not wearing the certification label, and they may still be able to provide what you want in the way of good green foods.

responsible. Interpretations of sustainable agriculture vary, but environmental stewardship balanced with economic profitability lies at the heart of the concept.

√  **Is It Grass Fed or Pasture Fed?** Choose grass- or pasture-farmed meats, eggs, and dairy when possible. Pasture-raised livestock (including dairy cows and beef cattle) avoids the illnesses from crowding in confined animal feeding operations (CAFOs) and promotes meat with a higher nutritional profile. Grass-feeding is environmentally friendly, relying on low-cost grasses that typically require little added water and few or no synthetic fertilizers and pesticides. The USDA requires that meats labeled as grass fed must come from animals fed solely on grasses, hay, and other nongrain vegetation. However, critics note that the USDA still allows animals to qualify even if they're confined to a pen and fed hay, and does not restrict use of hormones or antibiotics. (Third-party standardizations are being developed and may prove more stringent.)

√  **Is It American Humane Certified?** This is not a USDA program but rather one sponsored by the American Humane Association. Qualifying producers raise animals humanely in cage-free environments. This label on egg cartons indicates that the hens were raised cage-free (this was formerly known as the Free Farmed label).

The process is independently verified and focuses on animal well-being, not global warming issues, but sometimes the two overlap.

√  **Is It Seasonal?** That is, is it seasonal to the location where it's sold? Kiwi fruits ripen in April and May in Australia, but you can buy kiwis there all year round—grown and shipped from Italy. In the United States, most kiwi fruits hail from California and ripen in November, but because kiwi hold well in cold storage, the selling season of domestic kiwi stretches into April. So "seasonal" can be misleading; extended periods of cold storage create energy demands that recently picked produce do not. When shopping, consider what produce is truly seasonal, which are not seasonal to your locale at all, and which

## Scratching at Chickens and Eggs

Egg cartons can be billboards for greenwashing claims. *Free range* means nothing when it comes to eggs. This term, and its companion *free roaming*, applies to meat and poultry, and by USDA definition means the animal has had an opportunity for some exposure to the outdoors each day.

But it doesn't mean the animal actually did go outdoors, and it doesn't apply to eggs. As chickens go, a certified organic bird requires conditions that "provide for exercise and freedom of movement." Though the term *free range* sounds appealing, the USDA considers five minutes of open-air access per day adequate.

*Cage-free* can be misleading, too: it doesn't guarantee the animal's exposure to the outdoors and is not third-party verified; but American Humane Certified does verify true cage-free conditions, so look for this label. With USDA approval, "raised without antibiotics" is okay to mention, but it's not something that's verified by inspections or testing. And speaking of eggs, drop off empty egg cartons with a local egg farmer at your farmers market; reusing them in this way is more energy-efficient that just tossing them into a recycling bin.

come in shades of gray. (Wild fish are also seasonal, based on abundance and breeding cycles.) Your state's official agricultural Web site will have information on seasonal crops in your area, and many local farmers markets post what's ripe this week.

√ **Is It Local?** *Local* is a relative term, with the concern being both the distance a food travels to marketplace and the energy required to keep it fresh during transit (like refrigerated trucks). Some consider a radius of 100 miles local. Others find that 200 miles is more realistic. Farms, farmers markets, and some grocers sell local eggs, poultry, meats, dairy, honey, and vegetables. If you can find local organic and pasture-fed foods, then you'll be even greener. Think local, too, when you're buying wine, beer, and spirits; even if vineyards don't dot your landscape, small breweries probably do, and even vodka and tequila distillers are local to me in Texas. But some foods will never be local. When local's not an option (as in coffee, tea, or chocolate), seek out fair trade, organic, and sustainable options.

√ **Is It "Fair Trade"?** Fair trade is a market-based strategy for achieving both environmental and socioeconomic sustainability. It's intended to eliminate global poverty and promote environmentally sound practices by offering ways for farmers or other producers to make a decent living without exploiting the environment. (For example, instead of tearing down the rain forests for lumber, use the forests to shade organically grown coffee crops.) Various associations worldwide offer certification; in the United States and Canada, look for labels indicating Fair Trade Certified. Essentially, fair trade producers of goods (crops, livestock, and man-made items) are paid a fair price for sustainable products, with a further intent of building positive social impact. Fair trade principles promote fair prices, fair labor conditions (without forced child labor), ethical and direct trade (without middlemen if possible), reinvesting into community development, and farming sustainably without harmful chemicals. Underdeveloped nations have traditionally been the focus of the fair trade movement (with coffee, tea, chocolate, and spices among their leading crops), but domestic fair trade is a movement aimed at assisting small farms within the United States (growers of almonds and cranberries, for instance).

# Budgeting for Organic$

For many shoppers, the choice between organics and conventionally grown food is ultimately settled by price. Even the best-intentioned ecovore may be bound to household budget constraints. The cheapest way to enjoy organic produce is to buy in season, when supply is up.

To identify the most pesticide-prone produce, go to www.Foodnews.org, where the Environmental Working Group publishes a downloadable guide of specific produce with their average pesticide loads, called the Shopper's Guide to Pesticides in Produce. Or Google "organics on a budget" for other current resources.

But if you're on the run and can't always consult a guide, keep these simple guidelines in mind:

Some crops absorb more pesticide from below ground, or they have greater surface exposure to above-ground sprays. When budget means making choices, opt for organic:

- lettuce, spinach, and leafy greens

- root vegetables like potatoes and carrots

- produce with thin edible skins (like grapes, apples, berries, and peaches, and tomatoes and peppers).

Meats, poultry, and dairy products are costly because of organic grains and grass-feeding, but products solely free of hormones and antibiotics are more economical. Foods you peel before eating, like bananas and papayas, are less risky and are exposed to fewer pesticides.

When choosing produce, an ecovore might also consider:

*Is it globally plentiful?* Is it part of a global food shortage? How does it impact global food inflation? When biofuel grains ratchet up prices and are in short supply as global food grains, consider alternatives. Think potatoes instead of corn, for instance.

*Is it heirloom?* Heirloom seeds and crops help prevent cultivar extinction and promote biodiversity. For example, the Irish potato famine, which resulted in the deaths of more than a million people, was caused by a blight that decimated a single species. Disease resistance depends on biodiversity.

## The Hidden Cost of Cows

*"The daily food and water resources required to feed one cow exceed the daily amount of milk produced by that cow. A single cow produces 120 pounds of waste per day and improper disposal contributes to the pollution of underground water, in addition to nearby rivers and streams. The microbial fermentation that comes courtesy of a cow's digestive process produces as much as 100 gallons of methane per day. The methane emissions from one cow are higher than those produced by a car! California's Central Valley houses a majority of the state's dairy farms and, consequently, has some of the most polluted air in the nation."*—GreenLivingIdeas.com

## Meat: A Food Group with Issues

Whether to forgo eating meat can be a very personal choice. The most frequent reasons cited for giving up meat include improved health, animal welfare, religion, and, increasingly, concern for the planet. Too often, green arguments send mixed messages: they jumble all the reasons together. Granted, there are solid health and ethical arguments against eating meat, with plenty of emotional charge to them, but this is a book about planetary concerns. To stay on point, I've chosen to focus on one issue only: the direct ecological impact of meat.

### The Factory Farm Factor

Factory farming is the practice of raising animals in confinement at "high stocking density"—or more plainly, animals are crammed together, and for a portion of their lives receive no access to grass or vegetation. Also known as CAFO, concentrated animal feeding operations, they densely jam together enormous numbers of cattle, poultry, swine, and other species to achieve the highest output for the lowest cost. Today, most meats, poultry, dairy products, and eggs come from factory farms.

Factory farms have been spotlighted for their inhumane treatment of animals and for the health risks associated with feeding practices, infectious pathogens, growth hormones, overuse of antibiotics, and contamination through centralized processing—reasons enough to personally avoid factory-farmed meats, poultry, and eggs. (Think: mad cow disease, e. coli, and salmonella.)

But beyond the ethical and health issues, the environmental impact of factory farms is so huge that it affects the entire planet. It doesn't matter whether you live in Great Britain, the USA, or Madagascar. If you live on Earth, you're feeling the effects of factory farming, either directly or indirectly.

Of course, this doesn't mean that giant agribusiness is the sole contributor to environmental contamination. Any farm can pollute the environment, but the big guys have been doing it on a massive scale long enough for these

## The Centers for Disease Control's Statement on CAFOs

During the past three decades, animal production in the United States has become increasingly specialized. Many farms function as links in the chain of animal production, housing and feeding cattle and poultry. In 2003, the nation's 238,000 feeding operations produced 500 million tons of manure. The U.S. Environmental Protection Agency estimates that a small percentage of those facilities—called concentrated animal feeding operations (CAFOs)—accounted for more than half of the manure.

Public Health Concerns: People who work with livestock may develop adverse health effects, including chronic and acute respiratory illnesses and musculoskeletal injuries, and may be exposed to infections that travel from animals to humans. Residents in areas surrounding CAFOs report nuisances, such as odor and flies. In studies of CAFOs, CDC has shown that chemical and infectious compounds from swine and poultry waste are able to migrate into soil and water near CAFOs. Scientists do not yet know whether or how the migration of these compounds affects human health.

Pollutants possibly associated with manure-related discharges at CAFOs include:

*(continues)*

practices to dominate our food supply. With the added enticement of government subsidies, they sell their products at hard-to-resist low, affordable prices.

Small to mid-size producers are more apt, and perhaps more able, to raise foods in an environmentally friendly manner. Manure is just one example. Small-scale farms value and repurpose manure as fertilizer. The typical CAFO channels tons of manure into nitrogen-laden lagoons, which can pollute groundwater and waterways (with run-off into oceans as well) and essentially turn fertile land into cesspools. Anything that drinks or swims in manure-contaminated waters is affected. The manure lagoons also pollute the air with hydrogen sulfide, causing respiratory difficulties, burning eyes, headaches, diarrhea, and other problems to nearby residents and the animals

(The Centers for Disease Control's Statement on CAFOs, *continued*)

- *Antibiotics*, which may contribute to the development of antibiotic-resistant pathogens
- *Pathogens*, such as parasites, bacteria, and viruses, which can cause disease in animals and humans
- *Nutrients*, such as ammonia, nitrogen, and phosphorus, which can reduce oxygen in surface waters, encourage the growth of harmful algal blooms, and contaminate drinking-water sources
- *Pesticides and hormones*, which researchers have associated with hormone-related changes in fish
- *Solids*, such as feed and feathers, which can limit the growth of desirable aquatic plants in surface waters and protect disease-causing microorganisms
- *Trace elements*, such as arsenic and copper, which can contaminate surface waters and possibly harm human health

Researchers do not yet know whether or how these or other substances from CAFOs may affect human health. Therefore, CDC supports efforts to address these questions.

*Source:* Centers for Disease Control (CDC) and Prevention, U.S. Department of Health and Human Services.

themselves. Public and scientific pressure is being put on CAFOs to adopt more sustainable practices, though without economic incentives or penalties, they're not likely to change.

Fortunately, smaller producers with earth-friendly practices are finding a growing market share. Their products do cost more, partly because small farm operations don't receive the same government subsidies that CAFOs do and their fixed costs are naturally higher. But you can offset the price you pay at the store by eating less meat each week, which has the double-whammy effect of shrinking your cookprint by consuming less livestock overall.

## Making Choices: Consider the Meat or the Miles?

An enlightening report in *Environmental Science and Technology* (April 16, 2008), analyzed the environmental impact of red meat in the diet against eating local foods. The researchers assessed greenhouse gas ($CO_2$ equivalent) emissions in all stages of growing and transporting food. They found that:

- Switching to a totally local diet = driving 1,000 miles less per year.

- Replacing red meat and dairy with chicken, fish, or eggs for one day per week = driving 760 miles less per year.

- Switching to vegetables one day per week = driving 1,160 miles less per year.

- Switching from an average American diet to a vegetable-based one = driving 8,000 miles per year.

So, eating less red meat and dairy can be more effective in shrinking your cookprint than simply buying local food. But Carnegie Mellon University researchers Christopher L. Weber and H. Scott Matthews also point out that:

- Transportation creates 11 percent of an average U.S. household's greenhouse gases generated by food consumption.

- Agriculture and industrial emissions from growing and harvesting account for 83 percent of its greenhouse gases.

- The average distance traveled by food in the United States is 4,000–5,000 miles.

Of course, the ultimate cookprint-shrinking strategy includes both local food and foods lower on the food chain, but the researchers' numbers speak volumes about the types of impact we can make through multiple strategies.

## A World of Hurt: Healing Meat's Global Impact

Which causes more greenhouse gas emissions, rearing cattle or driving cars?

According to the United Nations Food and Agriculture Organization, the world's livestock sector generates more greenhouse gas emissions than does transportation (about 18 percent, as measured in $CO_2$ equivalent).

It's also a major source of land and water degradation.

The report calls for a drastic need to reduce livestock all across the planet. It notes in part:

- Global meat production will double from 229 million tons in 1999/2001 to 465 million ton in 2050, while milk output is set to climb from 580 to 1,043 million tons.
- Livestock are the main inland source of phosphorous and nitrogen contamination of the South China Sea, contributing to biodiversity loss in marine ecosystems.
- Meat and dairy animals now account for about 20 percent of all terrestrial animal biomass. Livestock's presence in vast tracts of land and its demand for feed crops also contribute to biodiversity loss; 15 out of 24 important ecosystem services are assessed as in decline, with livestock identified as a culprit.

The recommended industrial remedies range from improving livestock diets to recycling manure into fuel. But every person can make an impact, too. The most effective individual solution: eat less livestock and dairy.

*Shopping Guidelines: Pick Lower-Impact Meats and More*

Those among us who do eat meat would love to be able to buy meats with the greenest profiles. Sometimes, though, the best choices aren't fully available, or they may not fit your budget. When perfection's not possible, pick as many of these options as you can:

- Avoid CAFO meats, poultry, dairy, and egg products: almost all of these products in supermarkets come from CAFOs, but some stores

## The Food from Water Pipeline

The U.S. Geological Survey monitors water, among other environmental concerns. It lists these numbers as the amount of water needed to create a single serving of:

- Steak = 2,607 gallons
- Chicken = 408 gallons
- Milk = 65 gallons
- Rice = 36 gallons
- Almonds = 12 gallons
- Lettuce = 6 gallons
- French fries = 6 gallons
- Tomatoes = 3 gallons

Other ways to look at the numbers:

- A plant-based diet requires 300 gallons of water per day.
- A meat-based diet requires more than 4,000 gallons of water per day.
- If you eat one less beef meal each week, you'll save 40,600 gallons of water, 70,000 pounds of grain, and prevent the emission of 300,000 pounds of carbon dioxide each year (according to the Center for a New American Dream).

As droughts increase and water becomes more scarce, it seems that to conserve water, every glassful—and every bite—counts.

sell non-CAFO brands like Niman Farms, Berkshire Meats, and Organic Valley. Natural and specialty-food markets are more likely to sell non-CAFO items.

- Source out local suppliers: many small local farms sell directly to consumers, either at the farm or at farmers markets, or in local food stores.

- Choose chicken and fish over beef: the production of red meat is 150 percent more greenhouse-gas-intensive than chicken or fish. Even pork is less environmentally damaging than beef.

- Opt for organic, or at least grass fed, no antibiotics, no GMOs, no synthetic hormones, or all of the above. The more of these qualities the product has, the less impact it makes on the planet. Flip back to page 113 to for a guide to labels and what they mean.

## Green Cheese, Please

There's no denying it: some of the world's best cheeses come from outside the United States. But there's also no denying that some of the world's most inventive, tastiest cheeses are made right here within the USA, perhaps just down the road from where you live. California, Wisconsin, New York, and Vermont are the big cheese states, but more than 300 small cheesemaking farms churn out artisanal cheeses coast to coast. Search for nearby dairies on your state's agricultural Web sites or on Google.com. You might be surprised at what you find. The Mozzarella Company, for instance, makes award-winning cheeses just a few blocks from downtown Dallas. And while Texas may be cattle crazy, Austin's farmers markets and shops sell fresh goat cheeses that are local and organic. You can also pick up European-style cheeses, like local mozzarella in Maine, Gouda in Georgia, local feta in Florida, and local raclette in Ohio. For greener cheese, opt for products that come from non-CAFO dairies, and if not certified organic, ask if the cheeses are at least sourced from milk that's free of hormones, antibiotics, and preferably from grass-fed animals.

Though vegetarianism is a positive ethical and environmental choice, not everyone is ready to take that route, at least not completely. Admittedly, I'm an omnivore, but I've moved further away from animal-based foods with no regrets and great fulfillment. The recipes in this book are aimed at people who, like me, are willing to change their lifestyles for the good of the planet. They include meats, but as seen in the next chapter, they focus on scaling back portions and stretching the meat we eat. We can all make progress, even if we do it one bite at a time.

## Under the Sea: Fish and Shellfish

Scientists estimate that the oceans have soaked up about half of all carbon dioxide produced from fossil fuel emissions over the past 200 years. Had oceans not absorbed this carbon, current atmospheric carbon dioxide would be much higher than the current 381 parts-per-million (ppm)— probably closer to 500–600 ppm say climatologists.
—www.mongabay.com, article by Rhett A. Butler, March 8, 2007

Of all the food categories, fish and seafood opens up the biggest can of worms.

As a whole, the fishing industry is both a victim and a contributor to global warming. Some of the industry's problems are self-created—overfishing is one of them—but freshwater fish like salmon also suffer from hydroelectric dams and logging, and global pollution raises mercury levels in tuna. For the moment, let's push the issues of catching and eating fish aside to focus solely on the oceans themselves.

Mention the term $CO_2$ *emissions* and most of us think of atmospheric damage, to the air we breathe and the ozone layer. But the oceans are getting hit hard by fossil fuel emissions, and consequently, so is our seafood.

Releasing carbon dioxide into the atmosphere is the same as dumping industrial waste into the ocean. Oceans absorb one-third of the $CO_2$ generated by fossil fuels, and geophysicists predict that by midcentury, the oceans themselves will violate EPA water quality standards. When $CO_2$ dissolves in the seas, part of it transforms into the highly corrosive substance of carbonic acid. This acidification creates ph levels that effectively destroy the very creatures that form the foundation of the oceanic food chain.

Phytoplankton and zooplankton, for instance, don't appear as weekly specials in your grocer's seafood case, but without them, everything else you do buy (including salmon, mackerel, herring, and cod) is threatened. High acidity also corrodes corals and shellfish shells, and messes with oxygen supplies. The food-chain ripple extends from microplankton and corals to shellfish, fish, and mammals, and ultimately to humans. Researchers also find that the warming effect of temperature is bad, but right now it's far less destructive than the acidification of the seas. The solution: drastically reduce $CO_2$ emissions by switching to wind, wave, solar, and other alternative energy sources; and stop polluting the oceans with runoff from damaging land fertilizers, livestock, household cleansers, and automotive and industrial chemicals.

## What Seafood Is Best to Eat?

Shellfish and fish species are the hardest foods to recommend. Given the $CO_2$ emissions emitted by transportation, especially refrigerated containers, the choice to eat seafood at all, when it's not local, becomes even more complex. But mile for mile, fish still wins over beef. Fish in the diet can also bring salubrious benefits, good for the heart and the brain.

So if you do choose to eat food from oceans and waterways, which are the best choices? There's not an easy answer. Every day a new species seems to be in decline or threatened. Some species are so full of toxins from polluted waters or naturally occurring biotoxins, they're not safe to eat in quantity, and people with health issues should avoid them completely. With so many species in danger of depletion, some environmentalists encourage limiting consumption to two days a week. (Clearly, this is aimed at Japan and other nations with seafood-dominated diets, since meat still rules in the United States.)

The ecovore's policy: buy only currently safe and sustainable species (the list changes often). Check up on what's thriving and what's hurting at Seafood Watch on www.montereybayaquarium.org and at www.seafoodchoices.com. (Paul Johnson's award-winning book *Fish Forever* tackles sustainability and smart buying in great depth.) And get familiar with these guidelines:

### Sustainable Species
Choose species that are in abundance, and caught or raised without harm to the environment. Additionally, opt for short-lived species that reproduce

quickly; wild salmon, scallops, clams, oysters, and mahi mahi are good examples. Avoid "old growth" species, ones higher up on the predatory chain (like yellowfin tuna), and ones with slower reproduction rates (sharks, for instance, mature slowly and birth few offspring; they suffer more than other species from accidental catch and overfishing).

## Sustainable Fishing Methods and Aquaculture

How was it caught or cultured? Where? Choose wild species caught without harm to other creatures, coral reefs, or the environment (including aquatic habitats for birds, turtles, and fish "bycatch"). Select farmed species that are raised without spread of disease or pollutants, that don't rely on wild species as food, and that prevent farmed predator fish from escaping. Opt for fish from small-boat fishermen who practice sustainable methods, and look for sustainable retailing labels from the Marine Stewardship Council or Fish-Wise.

This is a complex area; it's hard to know if the fish you buy was caught or raised without harming the environment, other species, or yourself (with high mercury levels, PCBs, or biotoxins, for instance). But educate yourself on the issues, then open up a dialogue with fish merchants, and seek out resources to guide you in your current selections. Use your dollars to support fisheries that take innovative, proactive steps to improve their methods (such as trapdoors on shrimp nets to save turtles).

# EIGHT

# Energy-Efficient Ingredients: New Ways to Eat

What's for dinner? Sometimes you crave a certain food, other times you're open to suggestions: anything tasty and satisfying will do. So why not opt for ingredients that naturally cook with less fuel, or less water, or both? This chapter explores these types of ingredients, from no-cook pastas to nuts.

Energy-efficient ingredients also mean foods that put less stress on the planet, consuming less energy to raise or grow. Scaling back on meat and ramping up your intake of plant-based foods makes sense. But understanding *what* to do and knowing *how* to get there are two different beasts. This chapter offers practical solutions for balancing the world on your plate.

And to take these ideas off the page and onto the plate, the recipes in this book come with Green Meters, which point out ingredients with resource-reducing attributes and other green benefits.

## Energy Synergy

Quicker-cooking ingredients prepared with fuel-saving methods = energy synergy.

Besides raw food options, you may be surprised at how many hidden ways you can shrink your cookprint, simply by opting for whole ingredients that cook in little or no time—without tossing fast-food, shortcuts-in-a-box, or prepared foods in your basket.

Even similar ingredients (like dried peas and garbanzo beans) vary drastically in the cooking energy they require. A few surprising staples need no active cooking at all, including certain noodles and grains, and a presoak shortens cooking time for beans, rice, and other grains.

Also, dried grains, pastas, and beans are shelf-stable staples: they don't require cold transport or storage, and they last a long time (meaning fewer trips to the market). Tofu can be eaten with or without cooking, and when you buy it in shelf-stable aseptic containers, it also needs no refrigeration. These foods help stretch more resource-driven ingredients (meats, dairy, and vegetables) into complete meals.

The bottom line: besides choosing local and organic foods, you can control the very ingredients and foodstuffs that come into your kitchen. Certain ingredients, no matter where they come from or how they're raised or grown, are naturally more energy-efficient to prepare than others. And when you apply this book's Green Flame strategies, including using pressure cookers and rice cookers, you shrink your cookprint even further.

## Super Soakers

Some pastas and grains you drop into boiling water and boil until done. Others you pour water *over*, and walk away. In just a few flame-free minutes, they're soft, tender and ready to eat.

Click on an electric kettle for the ones that need boiled-water soaking, like bulgur wheat or couscous. With bean thread noodles and rice-paper wrappers, the water doesn't even need to be hot. These "super soakers" aren't just fuel-efficient, they're easy on the cook, require less equipment to wash, keep the kitchen cool, and are versatile enough to be part of a main dish or a stand-alone side dish. Consider these super soakers to get your creative juices flowing:

- **Bulgur Wheat** is the big boy behind tabbouli, a Middle Eastern grain salad (redolent with parsley, lemon juice, and tomatoes), but its range extends far beyond salads. Finely ground bulgur arrives at your kitchen almost ready to eat: just add water. It's essentially precooked at harvest then dried, an ancient means of preserving fresh wheat. Use it as an alternative to rice, serve it as a pilaf, mix it into meatloaf or meatballs, or season it as a salad with legumes, vegetables, and herbs. To cook, mix equal parts fine or medium bulgur and boiling water (and about 1/2 teaspoon salt per 1 cup of bulgur); let steep 10 minutes, then fluff with a fork. Coarse bulgur uses the same ratio of grain to water and salt but needs about 10 minutes of cooking on low, and a few minutes of fuel-free steaming after fluffing. A little bulgur goes a long way: fine and medium bulgur triple in size, and coarse bulgur doubles.

- **Couscous** is not a raw grain; it's a fine, bead-like semolina pasta, and the popular instant variety qualifies as a super soaker. Mix it with boiled water (or other liquid), cover, and let steam for 5 minutes, then fluff with a fork. Follow the manufacturer's directions, usually a ratio of 1 cup couscous to about 1-1/2 cups liquid, though this can vary. It comes in regular and whole-wheat varieties. Toasted Israeli or pearl couscous is peppercorn-size and requires a short amount of active cooking.

- **Rice Noodles and Rice Sticks** are made from rice flour and water, and range from wide ribbons to thin "vermicelli" to flat triangular flakes. Examples include Vietnamese banh pho and Chinese mi fen. Soak in hot tap water until tender: up to 30 minutes for wide ones, and 10–15 minutes for thin vermicelli. Drain and rinse in cool tap water. Mix into stir-fries, soups, braises, or stew-like sauces. Southeast Asians roll the thinner noodles into spring rolls with chopped vegetables, or pile them (thick or thin) into deep bowls with grilled meats, vegetables, and raw herbs on top. Pad Thai with Rice Noodles is a classic Thai dish (NewGreenBasics.com has the recipe). By the way, if you drop the dry noodles into hot oil, they puff up and explode into crunchy white clouds. (Another classic Thai dish, Thai Mee Krob, uses puffed rice noodles).

- **Bean Thread Noodles** are also known as cellophane noodles, or glass noodles because of their translucency; they're made from dried mung beans rather than grain flour, and they absorb liquids and sauces readily. For soft noodles, soak in water until tender, about 15 to 30 minutes. Drain. Use them like rice sticks, or try the recipe for Home-Style Glass Noodles on page 190.

- **Sweet Potato Noodles** have a wonderful chewiness with mild sweet flavor. Known as *dang myon* in Korean, they're the foundation of traditional Chap Chae, a braised meat and vegetable dish akin to sukiyaki. Soak in hot tap water until soft, 30 minutes or so, then drain.

- **Rice-Paper Wrappers are** translucent Southeast Asian sheets of rice flour and water that are brittle when dry, but become pliable with a quick soak. They come in round or triangle shapes. To prepare, soak one wrapper in water for 1 to 2 minutes (I pour water into a dinner plate or pie pan for this). Remove the wrapper when soft and drain on a towel briefly. Add a filling and roll up. They may be eaten cold and soft, with a dipping sauce. Or, roll them up tightly with the ends tucked in to seal in the filling completely, and fry in hot oil until crisp. Roll softened rice-paper wrappers around chicken salad, with extra lettuce, seasoned mayo and herbs, for a trendy East-West handroll.

Finally, check the label for made-in-the-USA pastas. Dried Asian noodles and wrappers are often imported, though they don't weigh much; but some are manufactured domestically, in the many Chinatowns and Little Saigon neighborhoods that stretch from California to New York, leaving a shorter transportation trail.

## No-Cook Pasta Sauces

Some classic Italian pasta sauces are totally fuel-free, or they make use of the cooking heat from the pasta pot. Consider these ideas:

- Over-the-Pasta Pot Sauce: place a large shallow bowl over the pot of pasta water, and let cheeses, cream, and/or other ingredients melt into a sublime sauce (no additional fuel required). After the noodles drain, toss them in the bowl with the sauce and serve. Example: Gorgonzola cheese and cream melted/warmed in the bowl, tossed with hot cooked ziti, and finished with chopped cilantro and toasted walnuts.
- No-Cook Tomato Sauce: chop summer's freshest tomatoes with garlic, green onion, and fresh basil; mix with olive oil and Parmesan and toss with pasta.

(continues)

## Good Things to Know About Grains

Adding more grains to the diet (and cutting back on animal proteins) makes good environmental sense, and nutritionists say they're healthier for us, too. Scoop them from bulk bins at most health and some conventional markets, which is also a good way to test a small amount to see what you like. Some dried grains need a lot of cooking to tenderize, but these energy-synergy tips cut back on fuel and time:

- Cook up a "grain bank," advises Lorna Sass, an expert on grains and pressure cookers. Whenever you cook grains, make an extra-large batch and save some, so they're always ready to reheat, toss frozen into soups, transform into entrees, or season as salads. Cooked grains last as long as five days under refrigeration and freeze remarkably well for months. With a grain bank, you'll spend less time in the kitchen and may find yourself eating less meat.

- Soaking most grains in water for 30 to 60 minutes shaves off as much as one-third the cooking time, and often reduces cooking liquid by 1/4 to 1/2 cup. Best candidates: wheat berries, hominy, farro, and rice.

---

(No-Cook Pasta Sauces, *continued*)
- Blanched Vegetables and Savory Oil: before cooking the pasta in the boiling water, blanch vegetables and toss them into a pasta bowl with olive oil, basil oil, or toasted hazelnut oil, garlic, and herbs; finish with Pecorino Romano or aged goat cheese.
Mix-Ins: deli-ready cured meats, olives, cheeses, and capers add easy zing to hot pasta. Bulk bins at Mediterranean markets (and some regular grocery stores) teem with robust brined olives, caperberries, peppers, and cured meats, ready to gussy up a sauce of fresh tomatoes or robust olive oil. Domestic versions can be just as tasty as imported ones, without a long-distance carbon trail. Look for specialty tuna from small Pacific Northwest canneries; Italian, Jewish, German, and Hungarian-style cured meats from California, Illinois, and Wisconsin; and cheeses from local and regional dairies.

## Speedy Little Lentils

Cooking times for dried legumes vary depending on age and type, but all things being equal, lentils are undisputedly the fastest cookers of all. Most are done in 30 to 40 minutes. If you soak them for 1 hour in plain tap water, they'll cook in about half the time.

Not bad, but this method cooks lentils using just 2 minutes of fuel (after the water boils): Combine 2 parts water with 1 part lentils. Cover and bring to a boil with dried herbs, like bay leaves or thyme. Boil gently 2 minutes. Turn off the heat and soak until tender. The soaking time is similar to conventional cooking time, but the method conserves energy, with less risk of overcooking. Drain the lentils, and season with salt, pepper, and olive oil.

- Starting grains in cold water, then bringing the water to a boil, results in softer grains; starting them in boiling water creates firmer grains.

- Pressure cookers make grain cookery super energy-efficient. The biggest hurdle with whole grains, especially, is the time it takes to soften the bran layer; pressure cookers speed up the process with tastier and nutritiously better results. Rice (even brown rice) cooks in 15 minutes, barley in 18, and rough-and-tough wheat berries mellow in 35 minutes.

Even without a pressure cooker, the more grains and the less meat you eat, the greener your cookprint will be. Books by Lorna Sass, Rebecca Wood, and even Betty Crocker provide complete recipes, tips, and formulas for cooking every type of grain, from amaranth to wild rice.

## Nuts to You! Nature's Ultraefficient Green Food

If you're cutting back on meat (or eat no meat at all), go nuts. Literally.

Earthwise, nuts are imminently sustainable: they grow on trees. And you don't need to uproot them to enjoy them. Wherever you live, there's likely a local nut growing nearby: pecans in the South, hazelnuts in the Pacific

Northwest, almonds in California, macadamias in Hawaii and Australia, walnuts from Canada to Argentina, and pine nuts from Colorado to Korea. Peanuts don't actually grow on trees: they're ground nuts, weird in that they grow at the end of tree limbs, in the ground.

Ounce for ounce, even shelled nuts are drastically cheaper than beef, both economically and ecologically. And they're more nutritious. The fats in nuts are healthy ones (good for the arteries, not bad like meat fats), and you don't need large portions of nuts to reap their nutritional or culinary benefits.

Eat them raw, toasted, or cooked into other foods. Keep containers of nuts handy to toss at will into, or onto, salads, pastas, cereal, breads, meat-loaves, pizzas, and desserts for intense flavor and irresistible crunch. And you'd be surprised what swapping one nut for another can do to change the flavor of a dish. Toasted pecan pesto, anyone?

When I started cutting back on meat, I found that nuts supplied an extra dimension of flavor, fat, and texture that kept me from feeling like I was chewing my cud, so to speak. With just a few nuts tossed into a dish, I felt sated and my body seemed to acknowledge it was getting the protein fix it wanted, too. Nuts aren't complete proteins, but they work with other foods to provide the essentials your body needs.

Nuts last for months—and even longer when refrigerated or frozen (to prevent rancidity). If you shell your own nuts, put the shells to use as mulch. Southerners grill with pecan shells as an aromatic alternative to wood chips. (Interestingly, ground walnut shells are mixed into textured paint and even oil drilling parts.)

So to shrink your cookprint tastefully and painlessly: go nuts.

# Putting It All Together:
## Strategies for Greening Your Meals

Realistically, if changing one's diet was an easy task, we'd all perform like athletes, have zero cholesterol, and look like those people on the covers of *Health* and *Fitness* magazines.

Whether it's a meaty or meatless path you take, going green is all about choices. Some choices are drastic and make huge differences. Some make

## Fuel-Efficient Flatbreads

Tortillas, pitas, naan, and lavosh are more than just convenient and versatile. They require less energy to bake, and because they're more compact, less energy to transport. Freeze them without taking up much freezer space. Locally made flatbreads are often sold in supermarkets, or look in ethnic markets (Mexican, Middle Eastern, and Indian stores especially). Warm them in a microwave or toaster oven, wrap or stuff them, and if they get dry or stale, slice into wedges and bake into crisp snacks next time the oven's already on.

less impact but are easier to assimilate. Bottom line: any green action is better than no action at all. And a bunch of little changes can make a difference collectively, especially concerning meat consumption.

### Getting Off the Meat Train

When I was a kid, mom served us balanced meals, but meat drove the train. For most families, it's still at the center of the plate, pulling ahead of the side dishes and leading the way when it comes to dominant flavors. The biggest step you can make in a cookprint-shrinking diet is to cut out or cut back on meat.

- If you choose to continue eating meat, eat less of it. Start by cutting down on portion sizes. Four ounces or less of protein is plenty for one serving (the size of a deck of cards). Opt for poultry and fish over beef, and add more plant-based foods to the plate.

- Make smaller portions seem larger. Slice grilled steaks, roasts, or chicken breasts into strips and feed four people instead of two, with heftier portions of side dishes.

- Treat meat as a seasoning, not the main course. Stir-fries, paella, curries, soups, and salads are delicious ways to stretch the flavor of meat, while leaning more heavily on vegetables, grains, and legumes as the body of the dish.

- Cook more meatless meals: if you usually eat meat every night, choose meatless entrees two nights of the week, and gradually increase their frequency to five nights or more. Ideas: risotto, pizza, stir-fried vegetables (with tofu, perhaps) and rice or noodles, quiche, and dishes rich in beans, vegetables, and grains.

- Keep things interesting: when cooking without or with less meat, emphasize variety, texture, color, and flavor to keep meals exciting. Going meatless doesn't mean being boring.

- Cook and freeze for creative options. Grains and dried beans take the same amount of effort to cook in large amounts as in smaller portions, and they freeze and reheat beautifully. Batch-cooking makes you more energy- and time-efficient, too.

- Go nuts. Nuts are powerful little nuggets of nutrition and good substitutes for meat and dairy. Instead of a small amount of meat in a stir-fry or other dish, toss in some toasted pecans, pine nuts, or peanuts. Vegan cookbooks transform nuts into healthy drinks, nut-loaves, and cream-like sauces.

- If you're already vegan or vegetarian, maybe the next step in your life is to organize meatless dinners, to inspire your friends and neighborhood carnivores to make their own delicious vegetarian fare.

Seriously, a taste is worth a thousand words. When you sample a cheese or tidbit in the market, you can tell for yourself whether it's something you want to eat again. More people might convert to meat-free diets if they could taste how good meatless meals can be. Part of adopting a new food or lifestyle is getting past the unknown. Every journey begins with a first step, or in this case, a first bite.

## Redesigning the Plate

Protein, starch, vegetable: it's the traditional dinner triangle of most American meals, where each element is served separately. It typically looks like this: steak, chicken breast, fish filet, or chop; accompanied by bread, pasta, rice, or potato; and another side dish of salad or cooked vegetable (like broccoli or yellow squash).

In many world cultures this type of meal would be a feast, and some would call it gluttony. Most of the world's population eats little or no fleshy food. But when you lift the lumps of protein off the Western world's dinner plate, you're left with elements that alone aren't always inviting or satisfying as a full meal. Unless you rethink the entire plate itself.

Traditional meals in Asia, Africa, Latin America, and other places emphasize grains, legumes, and, in some areas, dairy over animal proteins. Animal proteins, if served at all, are typically a luxury, enjoyed on special occasions or as seasonings in other dishes, or as a side dish themselves. Stews, too, stretch meager mouthfuls by being served with bread, grains, or pastas on the side, mixed in, or as a foundation; and beans and grains fill the pot more often than meat. Besides being more costly to raise, a quantity of livestock killed for food requires either a means of preservation (like curing) or cold storage, an additional energy hog. But not so with dried grains and legumes.

Try these twice a week: if your eyes glaze over at the thought of vegetable, grain, and/or legume dominated meals, take a look at some tasty dishes you may already be familiar with. These iconic examples can be served with or without small additions of meat, fish, fowl, eggs, or dairy, and the dishes may be rounded out with a solid portion of bread, noodles, or rice. A scan through some cookbooks or across the Web (try GlobalGourmet.com first) will yield a multitude of basic recipes for these classics:

**Italy and Greece**
Pasta e fagioli
Minestrone
Risotto
Polenta
Lasagne and pastas in general
Butternut squash ravioli
Eggplant Parmesan
Panini and pizza
Spinach and feta pie (spanako-
pita)

**France**
Crepes
Bouillabaisse
Cassoulet
Ratatouille

**Eastern Europe**
Kashka varnishkes

**India**
Curries
Dahls
Masala dosa
Biryani

**Middle East and North Africa**
Kebabs (meatless or mixed with
veggies)
Hummus
Falafel
Pilaf
Couscous
Tagine of vegetables and chick-
peas

## Asia

Stir-fry
Noodles with peanut sauce
Tofu, tempeh, and miso dishes
Sushi (with or without fish)
Donburi
Tempura vegetables on rice
Eggrolls, spring rolls, and won-
  tons

## South America, Mexico, and Spain

Bean-and-cheese burrito
Pozole with beans and chiles
Tamales
Enchiladas
Paella
Quesadillas
Black beans and rice
Quinoa, as a casserole or with
  sauce
Guacamole and chips or tortillas

## United States

Red beans and rice
Hoppin' John
Macaroni and cheese
Gumbo Z'herbe
Grilled cheese sandwiches
Chowder
Grits
Grilled vegetables
Hash

# NINE

# Waste Not:
# At Home,
# at Stores,
# in Restaurants

*To maintain a sustainable society, consumers must rethink their purchas-*
*ing and convenience expectations, as well as their material and energy*
*usage, to interact more intelligently with the world in which they live.*
*—Institute of Food Technologists*

At what point did single-serve containers become ubiquitous? Food packaging accounts for an alarming two-thirds of all packaging waste, which itself
makes up a third of all municipal solid waste (the rest comes from telephone
books, magazines, refrigerators, office paper, food scraps, yard clippings,
and other such sources).

Packaged-food convenience is like good and evil: it brings both freedom
and waste to our daily lives. In our craving for handy, time-saving convenience, we've let the packaging part of packaged goods get out of hand.
None of this mattered before consumers became environmentally chal-
lenged, but today, even big-box stores like Wal-Mart have vested interests
in more energy-friendly consumables.

Food waste is the other side of the equation, and we waste an alarming
amount of food, usually because foods spoil before we get a chance to use
them. Wasted food means wasting all the resources that went into growing
or raising it—including transportation fuel—and rotting food releases
methane, which is twenty times more harmful to the environment than $CO_2$.

When it comes to packaging, which are the best options? Once again, the greener path comes with gray areas, and the devil is in the details. This chapter captures the basics of reducing packaging waste and controlling your own food waste, and offers ideas for repurposing materials before they're sent off to recycling.

## Rock/Paper/Scissors: Glass/Paper/Metal—Which Wins?

Recycled and recyclable packaging are obviously better than conventional ones (provided you actually recycle). And the less packaging overall, the better the choice. (Skip the single-serving containers.) When it comes to balancing packaged products against environmental concerns:

- Glass beats canned or plastic

- Canned beats frozen

- Aseptic paperboard beats canned, plastic or frozen

Aseptic paperboard containers are juicy environmental alternatives (tomatoes, chicken broth, juice, milk, tofu, and even wine are packaged in them). They're recyclable, and they use far less energy to manufacture, fill, ship, and store products than comparable packaging materials. Their compact, lightweight brick design takes up less space, and their ability to preserve foods without refrigeration or preservatives saves energy.

Consider, too, powdered drinks instead of liquids, which take more fuel to transport because they're mostly water, or frozen cans, which need freezer trucks. If your family enjoys sports drinks and flavored beverages, Gatorade comes as a powder, as do lemonade mixes.

Scientists are actively developing new green materials, from bio-based films to the laminates and coatings on cans, paperboard, and plastics. The type of container you buy makes a big impact on marketplace. The Institute of Food Technologists' report *Food Packaging and Its Environmental Impact* (at www.ift.org) notes:

> Ultimately, the consumer plays a significant role in package design. Consumer desires drive product sales, and the package is a significant sales tool. Although a bulk glass bottle might be the best material for

fruit juice or a sports beverage, sales will be affected if competitors continue to use plastic to meet the consumer desire for a shatterproof, portable, single-serving container.

It's a good reminder that when you shop, your dollar makes a long-term impact on the environment in more ways than one. For a really clear explanation of packaging options, check out the Institute of Food Technologists' report, which includes a chart of materials, at NewGreenBasics.com.

To cut back on packaging:

- Bring your own bags or containers to the produce and bulk bins sections, and for carting your groceries home.

- Don't buy bottled water (it's no better than tap water, wastes fuel, and adds to the world's plastic pile). For bubbly water enthusiasts, Google home carbonation systems and DIY soda water for green alternatives.

- Skip single-serving containers.

- Opt for powdered mixes over trucked-in liquid drinks or frozen concentrates.

- Shop at the butcher counter (butcher paper is eco-friendly; foam and plastic wrap are not).

- Choose the brand that, volume for volume, uses less packaging.

- Pick the greener material (aseptic paperboard and glass over metal and plastic).

Most important, scale back your overall consumption to make the biggest dent in your cookprint. Even if you recycle, the truck that picks up curbside recycling spews carbon into the environment. Due to high fuel costs, some cities have ceased such programs entirely.

## Reusing Containers, Avoiding Disposables

Disposables made from recycled and recyclable materials are green options, but they're still a big waste. The greenest option is to reuse what you've already got before adding anything else to the landfill or recycling bin.

- Store leftovers in glass jars: prepared pasta sauce jars are as good as old-fashioned mason jars, and without their metal lids, they're microwave-safe. Spoon solids into them, too, not just liquids.

- Mix salad dressings for the week in glass jars, screw on the lid, and shake to blend.

- Use glass jars as organizing tools in the garage, workshop, and office areas. Store screws, nails, paper clips, rubber bands, pens, and pencils.

- Store bulk beans, grains, flour, sugar, granola, dried fruits, candies, and munchy snacks in glass jars or used plastic containers. Doggy biscuits, birdseed, and cat treats also work well in these containers, as do sesame seeds and spices—and the list goes on.

- Buying peanut butter in bulk? Repurpose a plastic container from home (pick one of similar weight to the ones provided at the store).

- Bring your own mug to the coffee stand or store; travel mugs come with lids and are insulated.

- Reuse plastic containers (like cottage cheese tubs) for storing leftovers, but don't microwave them; if you can't see what's inside, use a colorful grease pencil to label them as something other than what's printed on the container.

- Repurpose plastic containers as potted-plant saucers or cut drainage holes in them to use as pots; if they're unattractive, set them inside a cachepot or more attractive vessel (I line up three or four plastic-potted plants in a long window box, and if a plant dies, simply repot it without repotting the whole window box.)

- Repurpose a gently used length of foil (from covering a lasagna bake, or heating bread). Also: buy recycled aluminum foil; it's now available but you may have to Google a source.

- Skip disposable plates and glasses, and opt for durable glass or recycled plastic. Corelle is lightweight, made from glass, and comes with a three-year warranty against breakage; don't treat it like a Frisbee, but it's strong enough to resist breakage in most outdoor entertaining circumstances.

- Brown bagging it? Pack salads and sandwiches in repurposed plastic containers, and fill jars with beverages, soups, or salad dressings (seal tightly and make sure they don't leak). Pack cloth napkins, too.

Once I started buying in bulk, and bringing my own containers for bulk, produce, and grocery purchases, I noticed a drastic reduction in the amount of container waste in our household.

More traditional grocery stores are supplying bulk bin aisles, and where I live, Mexican supermarkets stock bulk bins with beans, grains, chiles, and spices—and they even offer butcher counters, seafood cases, and hot food delis. I've saved time and money many a night at the local Mexican market by having my containers filled with hot carnitas, cheesy enchiladas, freshly made flour tortillas, red rice, pinto beans, and fresh salsas, for a fraction of the cost of fast food or convenience items. The meal also tastes better, is made with fresh ingredients, and entails no packaging waste.

## Green Produce Bags and Ethylene Grabbers

In the old days, the only bags in the produce section were brown paper bags (which are still a good choice, especially for mushrooms). Today, how many rolls of plastic bags do we really need every time we shop? One bag for each type of vegetable? No. You can combine vegetables in one bag and let the checker weigh them separately when you pay. Better yet, bring and repurpose old plastic produce bags.

Or do what I do: bring green produce bags specially designed for longer storage life. Ethylene-absorbing bags really do keep fruits and vegetables fresher, and you can buy them in the produce section under various brand names (Evert-Green, BioFresh). Ethylene is a gas naturally released by produce, and it's responsible for ripening; trapped ethylene will hasten ripening (put hard peaches in a paper bag for a day to see what I mean), but when produce is already ripe, the gas speeds up decay and takes away freshness.

Regular plastic bags don't breathe, so they trap ethylene gas and hence shorten shelf-life. The ethylene-absorbing bags are reusable, so bring them with you, load your produce into them as you shop, and then plunk them directly into your refrigerator bins. You can also buy ethylene-gobbling discs and reusable freshness extenders and toss them into your fridge. One company, Fresh 'n Cold, makes a product that hangs by a suction cup in your refrigerator, absorbs gases, and is rejuvenated by sitting in the sun for a few hours. With the high cost of produce, especially organics, extending freshness makes good green sense.

## Storage: From Farm to Market to Home

A well-stocked pantry (including the refrigerator and freezer) reduces driving emissions caused by extra grocery trips, and shelf-stable foods make sound alternatives to produce shipped from overseas.

Dried foods lead the list in low-impact, shelf-stable pantry items—as in dried tomatoes, mushrooms, chives, cranberries, dates, figs, and raisins, for instance. Like nuts, they're flavorful and versatile, and most come ready to use, without cooking. They also cut a small transportation trail because they're lightweight, take up little space, and can be stored without refrigeration, both in-transit and after opening.

The downside of frozen foods is the volume of freezer-based emissions they create in every step of their journey. Consequently, shelf-stable goods in aseptic containers and cans are better options than frozen foods when speaking of fuel-efficiency. But the downside of both cans and frozen containers comes when they're lined with a substance known as BPA, which is addressed later in this chapter.

What about canned (or aseptic) versus fresh? Local fresh produce wins hands-down, but distantly grown produce changes the stakes. The definition of "distantly grown" can be hundreds or thousands of miles, but let's illustrate the issues with this example: "fresh" ears of corn may have been picked weeks ago, maintained in refrigerated warehouses, and transported for days in an emission-producing cold carrier before entering your grocery store for more cold storage. Fresh corn that's not sold gets thrown out as part of the store's waste. In contrast, canning facilities are often located close to the source, so the product gets canned while very fresh and needs no refrigeration. The can itself carries a manufacturing footprint, but it's recyclable. Fresh is expected to taste better than canned, but the quality of fresh produce can disappoint, too. When fresh foods travel, their canned counterparts often retain more nutrients. And increasingly, we're seeing more organics in the canned food aisle, at times and in places where fresh organic counterparts aren't available.

Now, consider canned versus frozen corn. Frozen corn usually comes in plastic packaging (or lined paperboard boxes). It undergoes the freezing process and must be maintained in a frozen state, from factory to railcar and/or truck, in the store, and in your kitchen. That's a whopping mess of damaging refrigerant material and fuel for the cold carriers, plus the transportation gases, spewing into the atmosphere. Sure, some frozen foods taste better than canned, but canned foods have come a long way in freshness. For most recipes, canned foods can be fine alternatives to frozen, especially given the ecological impact of each. Still, better packaging options exist, like glass and non-BPA containers, and new ones loom on the horizon.

## BPA: In Your Bottles, Cans, and Frozen Containers

As canned and frozen packaged foods go, BPA presents a real ecovore's dilemma. It's so ubiquitous, it's even in soda cans. From Con-Agra to Carnation, Annie's Naturals to Whole Foods, and conventional to organic, we've been eating products with BPA packaging for more than fifty years.

BPA, or bisphenol A, is a widely used chemical that can leach from packaging into foods and liquids. The Center for Science in the Public Interest stops short of putting all BPA-lined containers (including cans) on the do-not-use list. But it does note that pregnant women, fetuses, infants, and children are more at risk than the general population because BPA mimics estrogen, a hormone that affects brain development. In September 2008, a study published in the *Journal of the American Medical Association* found that adults with high levels of BPA in their urine were twice as likely to be diagnosed with heart disease or diabetes. Other studies suggest that as BPA leaches into ground water, it may harm fish and plants over time. (BPA does have a short half-life, chemically speaking, but it's everywhere; as a polycarbonate component, it's found in everything from CDs to medical equipment to fire retardant.)

In 2008 the Food and Drug Administration concluded that BPA-packaged products "are safe and that exposure levels to BPA from food contact materials, including for infants and children, are below those that may cause health effects. . . . At this time, FDA is not recommending that anyone discontinue using products that contain BPA while we continue our risk assessment process. However, concerned consumers should know that several alternatives to polycarbonate baby bottles exist, including glass baby bottles."

The food safety issues and related packaging remedies are really just opening up. Things you should know about BPA include:

- If you see #7 in the recycling symbol on a plastic bottle or frozen food container, it may contain BPA. But #7 is a catch-all category, so it includes both BPA and non-BPA containers.

- PVC containers marked as #3 can contain BPA in their plasticizers, but not all do.

- Any container of hard, clear plastic is likely to contain BPA, unless otherwise noted.

- BPA leaches out 55 times faster when exposed to hot liquids.

The good news is that non-BPA alternatives do exist. They're either not widespread or not promoted as BPA-free. For instance:

- The Eden brand uses non-BPA cans for beans (but not for tomatoes).

- Aseptic containers (as with tomatoes) and pouched packages (as with tuna) are non-BPA alternatives to cans (find more on aseptic packaging on page 142).

- For non-BPA plastic soda and water bottles, look for recycling symbols with 1 (PETE).

- Stainless steel and glass make good alternatives to hard plastic, polycarbonate bottles.

I anticipate that with increased consumer demand, more manufacturers will get the BPA out, which is another reason to voice your concerns and vote with your dollars. You'll probably never see labels stating the package contains BPA, but the brands that voluntarily go BPA-free will be smart to let us know.

So, what should you pick when stocking up on nonfresh foods? Opt first for shelf-stable packaging over frozen perishables; aseptic packaging beats canned goods. Canned goods can be better than frozen foods. But when cans contain BPA, the ecovore's solution may have to get down to personal choice, until clearer options come along.

# Clean Your Plate: Food Waste by the Numbers

Every time you throw food into the trash, you create methane. Methane, a major source of greenhouse gases, comes from landfills, among other sources. A new study is in the works, but according to a 1995 USDA report and recent numbers by other sources, food waste generates methane as well as these statistics:

- Americans generate 1 pound of food waste per day for every adult and child in the country.

- Americans throw out 27 percent of all food available for consumption (including food from homes, supermarkets, food service, and restaurants).

- British households waste a third of the food they purchase, and Swedish families with small children waste a quarter of theirs.

- Just 5 percent of the wasted food in the USA could feed four million people per day. Twenty million people could be fed on 25 percent of this nation's food waste.

- Americans generate 30 million tons of food waste annually, or 12 percent of the total waste stream, and 98 percent of the food waste ends up in landfills (not composted).

Most food waste consists of fresh produce, milk, grain products, and sweeteners. A family of four throws away 122 pounds of food each month. Of this, the heavy-weight waste shows up as:

24 pounds of fresh fruits and vegetables

22 pounds of fluid milk

18.5 pounds of grain goods (including bread and cereal)

10.5 pounds of processed fruits and vegetables (like canned and frozen goods)

15 pounds of sweeteners

10.4 pounds of meat and fish

8.6 pounds of fats and oils

What's being done about this? City Harvest in New York City drives to restaurants and picks up unserved food for charities and food banks. Composting programs in Massachusetts, San Francisco, and elsewhere turn food waste into compost for farmers (using produce and food not suitable for food banks, from supermarkets, restaurants, and homes). A Good Samaritan law protects donors from liability of food and groceries, encouraging more donations. Some restaurants are serving smaller portions, and a few college cafeterias skip trays, forcing students to eat only what they choose to carry.

The bottom line: buy food, eat it all, and don't let it go bad. Knowing how to store food properly and prolong its freshness helps keep waste and methane levels down. (www.Wastedfood.com has more practical tips for preventing food waste at home.)

## Unpacking the Groceries: Tips for Handling, Storing, and Thawing

Take good care of your foods, and they'll take care of you. After getting the goods home, or when storing leftovers or thawing foods, make these practices part of your permanent green routine to reduce waste and make foods last longer:

- **Store produce in nonplastic bags.** Ethylene-absorbing green produce bags really work, and so do low-tech cotton bags. Both are reusable and keep produce fresh longer than plastic bags. Bring them with you when you shop, too, to skip carting plastic produce bags home.

- **Store in glass with reusable lids.** When storing fresh foods and leftovers, pick glass over plastic storage containers. A number of companies (Boden, Anchor Hocking, Pyrex, for instance) make glass containers that go safely from refrigerator to microwave and oven, with either microwave-safe lids or oven-safe glass lids. Even when designed for food storage, plastic containers may emit hazardous gases, both when heated at home or in the manufacturing process.

- **Thaw foods in the refrigerator.** It's safer and more energy-efficient than thawing in a microwave. To prevent food-borne illnesses, don't thaw at room temperature unless recommended by the manufacturer.

- **Rewrap for oxygen-deprivation.** Oxygen hastens rancidity and staleness. If you buy big, repack a portion of the nuts, crackers, meats, or cheeses for later use. Seal them in reusable air-tight containers.

- **Store foods at their proper temperature.** Jump back to page 16 for ways to increase the chilling power of your refrigerator.

## Spending Your Dollars: Picking Greener Purveyors

Put the *eco* in economics by supporting businesses with greener products *and* greener practices. Today, the term *green grocer* goes beyond produce stands. Supermarkets are changing everything from freezer units to the vegetables they sell. Shrink your cookprint with every mile you drive, every package you buy, and even the restaurants you support. When it comes to picking a supermarket, don't just compare prices. Quality comes in many forms.

### *Warm Up to a Greener Chill in the Air*

Ever notice how much supermarket energy is devoted to cold foods—from freezers to closed-door refrigerators, to wide-open cases where cold air streams out? When a supermarket chain embraces greener practices, the impact can be huge but costly, and one worth supporting with shopping dollars. The EPA's GreenChill Advanced Refrigeration Partnership allies with supermarkets to reduce climate damage and increase efficiency. The results: fewer refrigerant leaks, extended shelf life of perishables, longer lasting equipment, and fewer emissions overall. Widespread adoption could cut refrigerant emissions by one million metric tons of carbon equivalent annually, *the equivalent of taking 800,000 automobiles off the road every year.* Early adopters include Whole Foods, Publix, Food Lion, Giant Eagle, Harris Teeter, and Hannaford supermarkets (for updates, visit www.epa.gov).

### *Go for Green Lights*

Does your supermarket use energy-efficient lighting? Does it leave the lights on even when the store is closed? Better yet, does it use solar or wind power? Consider these issues when choosing where to shop, and let merchants know you support green tactics, whether they practice them or not. (Think water, too: An HEB supermarket where I shop irrigates its landscaping from a rainwater collection system.)

## Seek Out Less Packaging and Local Foods

Does your market offer bulk bins? Bulk bins can be budget-friendly, and they cut back on packaging from the source to the market and from the market to your home. Buy grains, nuts, confections, cereal, and other items in bulk. More supermarkets are stocking local produce, eggs, and baked goods. Private-label products (i.e., the supermarket's own brand) may offer less packaging as well as organic options. Frequent the store's butcher case and seafood counter, where foods are wrapped in butcher paper without foam packaging. For larger amounts, invest in buying a part of an animal directly from a farmer.

## Support Specialty Shops and Markets

Local businesses often buy their ingredients locally, so it can make good green sense to shop at local bakeries, seafood and meat markets, produce stands, farmers markets, and even ice cream parlors that make their own sweet treats.

## Buy Direct

Join a community farm co-op and get a different seasonal selection of local produce every week, delivered to your door or neighborhood. Community-supported agriculture (CSA) farms are popping up everywhere, and you can usually choose to work the farm or not. Dairy farms often sell milk and cheeses directly to the customer, as do egg farms.

## Going Shopping? Change Your Habits

You can't regulate the cost of gasoline, but you can adopt driving and shopping practices that result in fuel conservation, fresher foods, and less food waste.

### Chill Out

Carry insulated bags and ice chests with freezer packs in the car so you can do a full day of errands at once, without food going bad.

### Cold-bag as You Go

As you shop, bag your perishables directly into the cold containers. The longer meats, eggs, and produce stay consistently cold, and the fewer times they reach room temperature, they longer they'll last overall. When you check out, they'll already be separated from the shelf-stable ones. Plus, no paper or plastic bags to bring home.

### Map Out Your Route the Right Way

Plan with efficiency in mind. As long as the perishables are kept cold and the frozen foods frozen, the groceries can be bought at any stage in the trip. (Purchase frozen foods every other trip, too, for more driving flexibility.) Take a tip from UPS: plan your route with right turns; left turns take more time and burn more fuel.

### Grow a Few Greens

Make fewer trips to market just to buy fresh produce. In our house, we buy the vegetables we don't grow, and we make salads from the ones we do. Anyone with a pot of soil and light can grow greens. Arugula sprouts like a weed in almost any condition (toss the seeds in soil, water well, walk away, and harvest in a few weeks, and every few days thereafter). Mesclun mixes mature in days. Even a small harvest added to store-bought lettuce helps flesh out the bowl and upgrades your salad from everyday to four-star quality instantly. (Plus, live plants bring their own carbon-cutting benefits.)

And finally, ride a bicycle or scooter. Shop nearby, and outfit your bike with baskets. In many countries, bicycles (and tricycles) are a dominant form of transportation, rigged with everything from wagons to sidecars. Sure, this is a bit more challenging as a suggestion, but it offers great appeal to the sports-minded and comes with fitness benefits to boot.

# Where You Dine

*I remember about four years ago, when a colleague asked me what was new and I said, we are into Waste Management. He replied, no, you can't say that.... You are into organic, delicious, local ... blah blah blah. I said, no, I am into waste management and my company is doing what it can to begin to minimize our waste. We now compost about 92% of our waste, recycle almost everything else, our garbage is now minimal and one restaurant is certified green.*

*—Jesse Ziff Cool, restaurateur/chef, author,*
*and sustainable foods advocate*

*Learning about sustainability is an ongoing process. . . . You can't do it all in one day. It's an education. You decide this week or this month we're going to try to tackle this. It might be something simple, like switching a cleaner. Once you've accomplished that, then you move on to the next thing.*

—Christopher Blobaum, chef (quoted in the Los Angeles Times)

Some chefs simply refuse to serve seafood, livestock, or produce that's not eco-friendly. And now a whole wave of restaurants, bakeries, and coffee houses are taking sustainability beyond the foods they serve, often with hefty start-up costs that deserve our green dollar support. With three successful restaurants in California's Menlo Park–Stanford area, chef-owner Jesse Ziff Cool has proven that green practices can also be profitable. She and other owners are greening up both the back of the house (the kitchen) and the front of the house (the dining room).

Visit an establishment's Web site to see if it broadcasts its green practices (many don't discuss, they just do it). And check out the Green Restaurant Association at www.dinegreen.com to learn more about what their certified green members are doing.

## How Restaurants Go Green: Low-Carbon Dining

Eating out impacts your cookprint just as much as cooking at home does. Choosing where you eat is therefore as meaningful as choosing what you eat. Even when you can't see the greenness on the walls, you may be surprised to find just how far an establishment goes to be green. Sandra Bullock's restaurant Bess, in Austin, Texas, for instance, even makes its own cleaning products from vinegar, baking soda, lemon juice, and salt.

Restaurants are going green by opting for:

- Biodegradable and nontoxic cleaning products

- Biodegradable disposable cups, flatware, and to-go containers (from sugarcane, for instance)

- Drinking-water-on-request policies

- In-house water filtration and carbonation systems, and not offering bottled water

## Trading White Linens for Green Savings

The next time you feel like treating yourself to dinner at a restaurant with white-linen tablecloths, think twice. Many establishments now are pulling away their tablecloths to reveal naked but attractive tabletops (or opting for placemats). According to Drew Nieporent, each of the four Nobu restaurants he co-owns worldwide saves $100,000 per year by not using tablecloths. Which also means saving on the power and water used to wash and press tablecloths, the materials and manufacturing resources to make the tablecloths, and the fuel used to transport materials, from the farm to factory, factory to wholesaler to distributor, distributor to restaurant, and back and forth from there on a regular basis to the cleaners. People go to Nobu for the food, and judging from the waiting list to get in, no one minds not eating on fancy tablecloths.

- Employee training sessions that promote and invite green solutions
- Energy-efficient lighting, and reduced lighting
- Fuel-friendly delivery vehicles (bicycles in urban centers, and hybrid cars)
- Induction cooktops
- Low-impact building materials, counters, tables, and chairs
- On-demand water heaters
- Organic and local food suppliers
- Produce and herbs from their own gardens, and eggs from their own chickens
- Recycled, unbleached paper products
- Recycling cooking oil into biodiesel fuel
- Repurposing cartons and boxes as take-out containers

- Restrooms with sensor-activated faucets and dual-flush toilets
- Sending organic waste to local farms for composting
- Solar and renewable energy, often built on-site
- Sustainable and seasonal food menu items (and avoidance of nonsustainables)
- Water-conserving dishwashing equipment and targeted kitchen policies

On a related note, some college cafeterias no longer provide trays; consequently, students take fewer items and waste less food, and the policy eliminates wash water, special tray handling equipment, and the manufacturing footprint of the trays themselves. Schools of all levels are building campus gardens into their dining room food chain, too.

# Saving the Planet:
# One Cook at a Time

*I started to walk down the street when I heard a voice saying: "Good evening, Mr. Dowd." I turned, and there was this big white rabbit leaning against a lamp-post. Well, I thought nothing of that, because when you've lived in a town as long as I've lived in this one, you get used to the fact that everybody knows your name. And naturally, I went over to chat with him.*

   *—Jimmy Stewart explaining the pooka, his magical pal, in* Harvey *(1950)*

An interesting thing happened as I was writing this book. I started practicing my own advice.

Besides seeing measurable results (like reducing packaging and food waste), I ran into an unexpected outcome. I became more connected. Ways to go greener started popping up *everywhere*.

Like the imaginary rabbit in *Harvey*, they invaded my consciousness and won't go away.

Big green whiskers keep tickling my thoughts. Not just in the kitchen or the home, but in every store purchase, mile driven, vote cast, and dollar spent.

The reversal of climate change requires a complete paradigm shift and global actions, in more than just food and cooking. But one thing leads to another. Little steps in behavior can make a big difference in how we think.

Will you adopt every practice in this book? Probably not.

Will you change some of your habits? Probably so.

Will you become aware of your actions, and those of others? Absolutely.

This book captures a moment in time. Use it as a guide, and keep revising it. Fixing our mess is the feel-good meme of the moment, but we need long-term commitment. Technology, attitudes, and research keep churning out green breakthroughs and new consumer "do's and don'ts." (Remember when you first heard about the goodness of compact fluorescent bulbs, but no one mentioned the mercury issue? Information gaps happen.)

Go to NewGreenBasics.com for the references, research, and unresolved questions behind this book. Blog-on with personal tips, recipes, challenges, and facts. Go guerilla: take this book to the water cooler and start a *"Did you know . . ."* discussion. Host a dinner party cooked with New Green Basics methods, using the recipes that follow. Powerful ideas migrate, if you share them.

And if this book conjures up your own green *Harvey*, don't ignore him. Welcome him to the table, pour him a glass of organic wine, and start chatting. Who knows what you'll cook up? A six-foot rabbit may be just what we need to raise our green-consciousness in the kitchen and beyond. After all, everybody eats.

Even pookas.

# 40+ Ways to a Cookprint-Shrinking Consciousness

Reduce, reuse, recycle, repurpose, and *rethink* how you cook, shop, and eat.

*To shrink > think:*

Energy-efficient kitchen zones

Water conservation and reuse

Lower hot-water usage and
   temperature

Energy-Star appliances

Small appliances as fuel-savers

Electric teapots over cooktop boiling

Avoiding peak power hours

Unplugging appliances

Renewable energy sources

Lower-emission grilling

Nontoxic, biodegradable cleansers

Regular over antibiotic cleansers

Reusable cloth napkins

Recycled and recyclable products

Plants over animals

Non-CAFO products

Local

Organic

Seasonal

Sustainable

Energy-efficient ingredients

Weather-sensitive cooking

Cooktop before oven

Induction burners

Passive cooking over active fuel use

Skipping the preheat when possible

Toaster ovens

Convection cooking

Microwave cooking

Simultaneous baking

Multitasking boiling water

Fuel-efficient cookware

Nontoxic cookware

Farmers markets and direct from farms

Fewer grocery trips

Shelf-stable over frozen

Minimal packaging

Bulk-buying

Aseptic and glass over cans

Extending food storage

No food waste

Green-conscious grocery stores

Low-carbon restaurants

# THE NEW GREEN BASICS RECIPES

# About the Recipes: Cooking the New Green Basics Way

To make it easier for cooks to start putting fuel- and water-saving methods into practice, this collection spotlights familiar dishes, common ingredients, and everyday fare. Before long, you'll unlearn old ways, and greener methods will become second nature.

## Read the Green Meters

To shrink your cookprint:

• Eat lower on the food chain, and

• Cook with less fuel and water.

Green Meters identify the areas where each recipe registers highest on the low-impact scale. The Green Meter also helps cooks gauge how long the dish takes to make, ways to adapt it, when its ingredients are in peak season, and any special conveniences. The last bullet in the Green Meter points out how you can convert the old habits of conventional cooking into your New Green Basics repertoire.

Remember: for those who eat meat, downsize the portions and stretch the servings, as these recipes suggest. Whether meat is the main event or used as a seasoning, consider substituting meat-free alternatives. Tofu, tempeh, and seitan are popular vegetarian staples, but some folks find them an acquired taste. Plenty of natural ingredients that are full of flavor and texture can replace meat, fish, or fowl—like meaty-tasting portobello mushrooms, sun-dried tomatoes, black beans or chickpeas, dried fruits, pecans or walnuts or other nuts, sunflower or pumpkin seeds, corn or hominy, and feta or cheddar, just to name a few. As a reminder, flip back to Chapter 7 for tips on sourcing non-CAFO meats, poultry, cheese, and eggs—another big step in shrinking your cookprint.

Most recipes in Meat-Free Mains 'n' Grains and in Vegetable Sides and Salads are substantial enough to serve as meatless main courses. Consider them as stand-alone components of a low-carbon diet.

When cooking the recipes, keep these general pointers in mind:

- Whenever you can repurpose water, please do. Salted water has limited uses (don't pour it on houseplants, for instance), but it's fine for prerinsing messy pans or soaking food containers.

## Safety Note: Take the Temperature

With all cooking, check for a safe internal temperature. Allow standing time, which completes cooking, when checking the internal temperature with an instant-read food thermometer; "carry-over cooking" can raise internal temperature 5 to 10 degrees. The USDA has revised previous standards and now recommends these temperatures:

- *Beef, veal, and lamb steaks, roasts, and chops to 145°F.*
- *All cuts of pork to 160°F.*
- *Ground beef, veal, and lamb to 160°F.*
- *Egg dishes, casseroles to 160°F.*
- *Leftovers and stuffing (separate from the bird) to 165°F.*
- *Poultry, white and dark meat, should now reach 165°F. (The old standard was 170 for light and 180 for dark.)*

Finally, visit and revisit NewGreenBasics.com for updated strategies, recipes, and lifestyle tweaks to keep shrinking your cookprint.

- Potatoes take more time to cook than most foods, so you'll find several fuel-efficient methods demonstrated in these recipes. Green beans aren't so time-consuming, but you'll also find more than one way to tackle them. By having a variety of options, you're better able to piggyback, cook simultaneously, or opt for other ways to stretch fuel and water. Even quicker-cooking foods like snow peas or fish can find room for improvement, by saving energy or water, or both. The headnotes explain the hows and whys, to help you cook flexibly and adapt techniques for even greener results.

- Read the headnotes at the top of recipes, especially with these recipes. They've got important tips for making the dish, and for making it greener. As a practical reminder: read recipes all the way through before deciding to make them, so you don't stumble onto surprises.

- Recipe prep and cook times are approximate, based on having the components at hand. Rummaging around in the fridge or pantry to find ingredients is not included in prep time.

- Any recipe takes longer the first time you make it. Allow extra minutes to reread and check the directions as you go. After you've seen the results and know what to expect, cooking always goes faster.

- Your specific fuel, cookware, and ingredients all impact cooking times. Recipe clues for doneness (like when it's "golden brown") are better indicators than the specified minutes in the oven or on the stove. Cook flexibly, taste for seasonings, and adjust accordingly.

# Small Bites, Appetizers, and Soup

## MEDITERRANEAN RICE-PAPER ROLLS
Makes 6 (3 entrees or 6 appetizers)

### Green Meter

- Green Goodness: No-cook; shelf-stable wrapper conserves fuel
- Prep/Cooking Times: 15 minutes, start to finish
- Prime Season: Year-round
- Conveniences: Make-ahead, versatile option for brown-bag lunches or entertaining
- New Green Basic: Use instead of sandwiches made from baked bread; can be meat-free, too.

*Don't limit rice-paper wrappers to Asian ingredients or dipping sauces. In this recipe, the wrappers are tightly rolled around strips of chicken, a lemon-caper sauce, shaved parmesan, crunchy vegetables, and lettuce. Soaking the wrappers in seasoned rice vinegar infuses them with a bright, perky flavor, making messy dipping sauces unnecessary. They're neat to eat, can be prepared in advance, and travel well as brown-bag lunches or picnic fare. Cut them in half for cocktail fare.*

*New Green cooks can use rice-paper wrappers as a shelf-stable, no-cook alternative to breads or tortillas; they never need refrigeration and last indefinitely. Other filling ideas? Sustainable shrimp and grilled fish, tofu, hummus, an infinite assortment of greens, herbs, pickles, and fresh or blanched vegetables, like snow peas, cucumber, jicama, olives, radish, chiles, and chives.*

Note: *Seasoned rice vinegar is plain rice vinegar to which some sugar has been added. It's available in the vinegar or Asian aisle. Buy packages of rice-paper rounds where Southeast Asian products are sold; if you use 6-inch wrappers, distribute the fillings accordingly.*

Sauce:
1/4 cup mayonnaise
1 tablespoon fresh lemon juice
1 tablespoon chopped capers
1/4 teaspoon black pepper

Fillings:
1/2 bell pepper (green, red, or as desired)
1 carrot
1 rib celery
2 green onions
6 small to medium red leaf or other lettuce leaves
8 ounces cooked, shredded chicken meat (1 large breast)
1 ounce shaved Parmesan

Wrappers:
1/2 cup seasoned rice vinegar plus 2 tablespoons water
6 (8-1/2 inch) rice-paper wrappers

1. Mix together the mayonnaise, lemon juice, capers, and pepper for the sauce. Slice the bell pepper, carrot, celery, and green onions into matchstick-size pieces, or narrow and thin strips of 2-inch lengths. (Use the green and white parts of the green onions, and include any celery leaves.)

2. Fill a shallow plate or pan with the seasoned rice vinegar and water. Soak one wrapper at a time until pliable (a minute or so). Lift up the wrapper and let excess

liquid drain back into the plate. Smooth out the wrapper on the work surface. As you remove one wrapper, replace it with another one to soak.

3. To assemble one roll: In the lower third of the wrapper, place a lettuce leaf so the frilly part partially extends over the left or right edge. Distribute a few strips of chicken over it. Drizzle on a small amount of sauce. Add strips of bell pepper, carrot, celery, and green onion and Parmesan shavings. Gently pull the wrapper's bottom edge up and over the filling and tighten by pulling back slightly. Give the roll a forward turn. Fold the end without the lettuce in, and continue rolling the wrapper into a cigar, or if the fillings are bulky, a cone shape. Place seam-side down and continue making the remaining rolls. Cover and refrigerate until ready to use. The wraps will keep up to 6 hours. *(Note:* You can also fold in both ends of the wrapper to completely contain the filling.)

# HERBED SALMON GRAVLAX

Makes 20 small servings; may be halved

## Green Meter

- Green Goodness: No fuel or water; good use of sustainable wild salmon
- Prep/Cooking Times: 15 minutes prep + 5 minutes per day for 2 days
- Prime Season: Year-round, but wild salmon season runs May through September
- Conveniences: No hot cooking; lasts up to a week; ready to eat; easy entertaining
- New Green Basic: Use instead of cooked salmon; serve small portions to stretch servings.

*Swedish gravlax is salmon cured with no heat at all. It tastes different from lox or smoked salmon, but the results are equally as spectacular— and the vivid color on a huge platter adds festive glamour to a table. Essentially, you make gravlax by coating fresh salmon fillets with salt, sugar, and herbs, weighting the fish down, and refrigerating until cured.*

*When ready, a slice of gravlax is not salty, dry, or raw—it's just a delicate sliver of heaven.*

*Serve gravlax as a centerpiece main course, with lemon wedges and caper mayonnaise, or as an appetizer on small breads, with radish slices and creamy horseradish sauce or garlic-Gorgonzola puree. It's a four-star sandwich made with cucumber slices, goat cheese, and watercress or tomato, and an easy breakfast with cream cheese, bagels, red onion, and lemon, or scrambled in eggs, with avocados, sour cream, and chives.*

Servings: *To halve the recipe using 2–3 pounds salmon, cut a single whole salmon fillet across the middle, and lay one half over the other (thick ends over thin ends). Cut the ingredients in half, but don't cut the curing time—you'll still need 48 hours.*

2 whole fresh salmon fillets with skin, about 5 pounds total
3/4 cup kosher or coarse salt
3/4 cup sugar
1 large bunch fresh herbs, such as dill, basil, or cilantro, or a mix (2
  to 3 cups when chopped)

1. Run your fingers down the fillets and feel for small rigid bones. Using needle-nosed pliers or tweezers, pull these pin bones from the salmon. Mix together the salt and sugar in a small bowl. Coarsely chop the herbs.

2. In a large glass baking dish or other nonreactive container, distribute one-fourth of the salt-sugar mixture in an area just the size of the fillet. Place one fillet skin-side down on the mixture. Sprinkle two-thirds of the remaining mixture evenly on top of the fillet, along with all of the chopped herbs. Place the second fillet flesh-side down on top of the herbs and curing mixture. Sprinkle the remaining curing mixture on the skin side of the second fillet.

3. Weight the fillets: set a plate, pan, or small cutting board on top of the fillets (it should be large enough to cover them but small enough to fit within the dish). Add something heavy for weight: a cast-iron pan or a pot filled with water, or wrap some bricks in foil.

4. Refrigerate 2 to 3 days, turning the fish twice a day. The salmon releases juices as it cures, so be careful when moving the pan. (Pour off accumulated juices if the pan gets too full.) Scrape away excess cure. To serve, thinly slice the fish at an angle. Gravlax will keep a week or more refrigerated.

Variations: *I never make gravlax the same way twice. Sometimes I add some white or black pepper, a splash of vodka, or mix cilantro and basil, or dill and basil. I've even mixed brown sugar and white sugar together, for a more caramelized taste. But don't go too crazy; you want the delicate salmon flavor to shine through, not be overwhelmed.*

## FRIJOLE FUNDIDO, RAPIDO
Serves 2 as a main meal, 4 as a snack

### Green Meter

- Green Goodness: Microwave replaces oven, saves fuel; kitchen stays cool; meat-free
- Prep/Cooking Times: 5 minutes prep + 5 minutes cooking
- Prime Season: Year-round
- Conveniences: Quick 'n' easy, one-dish cooking; pantry standby
- New Green Basic: Use beans and vegetables to stretch cheese; microwave to conserve fuel.

*For a meal or a snack, Queso Fundido couldn't be easier. Translated as "melted cheese," it's simply a hunk of cheese baked in an earthenware casserole for about 15 minutes in the oven, until melted. Once bubbly, the cheese is served communally, for all to spoon into warm tortillas. It's absolutely delicious.*

*This microwave version saves fuel and captures the same luscious flavor in a more balanced dish, by baking creamy frijoles with onion, bell pepper, and a dash of cumin under a gooey layer of cheese, then topping it with fresh tomatoes. Use any good melting cheese (a locally made favorite, perhaps?) or domestic Mexican-style queso, Monterey Jack (plain or spiced), sharp cheddar, fontina, or mix a mellow cheese with a nutty one (like Jack and Gouda). Thaw out some beans you've cooked and frozen, or grab a pantry staple, like Ranch-Style pintos, Southwestern seasoned beans, or black beans. Add more vegetables, like corn, zucchini, celery, or* roasted green chiles. Kick up the spices, too, with some chile powder, Tabasco, or salsa.

Got kids? Let little hands help with this no-flame, nonsweet dish. They can grate cheese, measure ingredients, and stir, and older ones can chop vegetables. It's delicious and nutritious.

> 1 teaspoon olive oil
> 1/2 teaspoon ground cumin
> 1/2 medium onion, chopped
> 1-1/2 cups cooked pinto beans (home-cooked, or a 15-ounce can, drained)
> 1/2 medium bell pepper, chopped
> 8 ounces (2 cups) shredded melting cheese (like Monterey Jack, sharp cheddar, or Gouda)
> 1 ripe tomato, chopped
> Warm flour and/or corn tortillas (see Note)
> Garnishes: shredded lettuce; salsa (optional)

1. In a microwave-safe casserole with lid ajar, microwave the olive oil, cumin, and onion on high until the onion softens, about 1 minute.

2. Stir in the beans. Top with the bell pepper and cheese. Microwave on high, lid ajar, until the cheese melts, about 2 minutes.

3. Sprinkle with chopped tomato and serve with warm corn or flour tortillas. Let each person spoon the filling into a tortilla, adding shredded lettuce or salsa if desired.

Note: *To heat tortillas, stack on a microwave-safe plate, cover with a folded dishtowel, and heat on medium high 1–2 minutes. For even heating, rearrange the tortillas in the stack halfway through, so the middle ones are on the top and bottom.*

# VIETNAMESE GLASS NOODLES SOUP WITH CHICKEN
Serves 2 as a light lunch or dinner, 4 as a starter

## Green Meter

- Green Goodness: No-cook noodles conserve fuel; short cooktop time for soup
- Prep/Cooking Times: 20 minutes prep, mostly unattended + 7 minutes cooking
- Prime Season: Year-round
- Conveniences: One-pot dish; uses precooked chicken; little chopping; few ingredients
- New Green Basic: Use no-cook noodles instead of boiled pasta.

*Bean thread noodles "cook" just by soaking in tap water. Made from mung beans; they're sold as opaque white, brittle noodles. When soft, they turn silky and translucent, hence the name "glass noodles" or "cellophane noodles." Found in Asian markets and some supermarkets, bean thread noodles come in small nests of about 1 ounce, usually bundled with 6 to 10 nests per package, or in larger bundles. Bean thread noodles suck up sauce and broth like mad, so add enough of them and they'll transform a brothy soup into a full-fledged noodle dish, if you prefer.*

2 (1.32-ounce or approximate) bundles bean thread noodles
10 dried cloud ear or shiitake mushrooms, or a mix
1 quart chicken broth (low-sodium preferred)
1-1/2 teaspoons sugar
1-1/2 tablespoons fish sauce (nam pla or nuoc mam), or to taste
Salt
1 cup packed shredded, cooked chicken meat
2 green onions, green and white parts sliced thin
1/4 cup coarsely chopped cilantro
1 large lime, quartered
1 minced small green chile, or dried red chile flakes to taste

1. Soak noodles in tap water (tepid or hot) to cover until soft, about 20 minutes. Drain and snip with scissors into shorter lengths (about 3 to 4 inches). Soak the

mushrooms in hot tap water to cover until soft, about 15 minutes. Rinse to remove any grit, squeeze dry, and snip into bite-size strips, discarding tough stems.

2. Bring the broth to a boil. Stir in the sugar, fish sauce, mushrooms, noodles, and salt to taste. Turn off the heat and let set a few minutes to warm all ingredients. Ladle into bowls. Sprinkle on shredded chicken, green onion, and cilantro. Serve with lime wedges and chile on the side.

# Main Courses with Meat, Fish, or Fowl

## STIR-FRIED VIETNAMESE CHICKEN WITH PASSIVELY BLANCHED SNOW PEAS

Serves 4 with rice

### Green Meter

- Green Goodness: Stir-fry and passive blanching conserve fuel; rice stretches protein
- Prep/Cooking Times: 7 minutes prep + 2 hours or less marinating + 5 minutes cooking
- Prime Seasons: Spring (if using snow peas); year-round for chicken
- Conveniences: Can serve with reheated rice; little chopping or mess; quick 'n' easy
- New Green Basic: Use as a template for low-prep stir-fries; use the microwave passive-blanching method for other vegetables.

*Unlike other stir-fries, this dish requires little chopping, and the snow peas need none. To speed up prep, stack two chicken pieces on top of each other and slice them at the same time, using a sharp chef's knife or cleaver.*

*Even when stir-fried, snow peas taste better after a quick passive blanch (which may be done in advance), to take out the rawness and keep them bright. This dish yields a lot of juices and is meant to be served as a rice-bowl dish or over a mound of rice on the plate. Short-grain brown rice is especially tasty here. Fish sauce, known as nuoc mam in Vietnamese and nam pla in Thai, is sold in Asian aisles and adds the necessary salty balance.*

1 large lemon, washed and dried
3 cloves garlic, minced
3 tablespoons fish sauce (nuoc mam or nam pla)
2 tablespoons soy sauce
1 teaspoon finely ground black pepper
Dash red pepper flakes
2 pounds boneless, skinless chicken thighs or breasts
3 tablespoons molasses
2 tablespoons dry sherry
1 tablespoon canola or vegetable oil
6 ounces snow peas, passively blanched (see below)
Steamed brown or white rice for serving (about 6 cups cooked)

1. Finely grate the lemon's zest into a mixing bowl. Juice the lemon (about 3 tablespoons juice) into the bowl. Add the garlic, fish sauce, soy sauce, black pepper, and red pepper flakes. Cut the chicken into finger-thick strips, compatible with the size of the snow peas. Toss the chicken in the marinade and let rest 30 minutes or preferably 2 hours, refrigerated.

2. While the chicken marinates, mix together the molasses and sherry.

3. Heat the oil in a wok or large skillet over high heat until almost smoking. Swirl the oil in the pan. With a slotted spoon or spider, carefully add the chicken, leaving the marinade in the bowl. Stir-fry until the chicken is cooked on the outside but still raw in the center. Stir in the molasses mixture. When the chicken is almost cooked, pour in the marinade from the bowl and cook another minute or so, until the chicken and juices are cooked through. Turn off the heat. Mix in the snow peas. Serve hot over cooked rice.

**Passively Blanched Snow Peas:** Place 6 ounces trimmed snow peas in a heatproof glass baking dish or Pyrex bowl. Cover with boiling water (about 2 cups). Let soak 1 minute, or until just tender. Scoop out the peas, place in cold tap water until cool, and drain. Peas may be prepared 2 hours in advance and left at room temperature. Add to a stir-fry at the end of cooking, serve as a side with sesame oil and soy sauce, or chill overnight and serve in a salad. (Repurpose the blanching and chilling water.)

## GINGER CHICKEN AND BROTH, PASSIVELY POACHED

Serves 4 to 6

### Green Meter

- Green Goodness: 20 minutes of cooktop fuel cooks a whole chicken, with healthful broth
- Prep/Cooking Times: 5 minutes prep + 15 minutes active cooking + 60 minutes unattended
- Prime Season: Year-round
- Conveniences: Can be made in advance; will keep 3–4 days. Morphable: shred meat into soup, salads, sandwiches, tortillas, or serve pieces with soy sauce and rice. Use the broth wherever chicken broth is called for.
- New Green Basic: Use this method for any chicken soup or simple cooked chicken meat. Vary with chile and cumin for a Mexican spin, or saffron and curry powder for a Bombay boost.

*In this Chinese poaching method, chicken simmers for a short time immersed in seasoned water, then the flame is turned off and the chicken continues to passively cook in the hot liquid. The meat is more tender and moist than if boiled. My favorite cold remedy? A bowl of this chicken and broth, with noodles, and bok choy or shredded Chinese cabbage. Even without a cold, you'll feel healthier.*

*If you cook this dish in advance and refrigerate it, the fat hardens and lifts off. Also, the chicken meat stays moist in the broth and is easier to shred when cool. I usually stretch this dish across three meals: first, as pieces with a sesame-soy or Vietnamese dipping sauce and rice. Next, the broth becomes wonton or noodle soup. Finally, the remaining chicken shreds into Hazelnut Chicken Salad on Shredded Napa Cabbage (see recipe below).*

1 whole chicken (about 4 pounds), without giblets
2-3/4 teaspoons salt
1-1/2 inches fresh ginger, cut into 6 slices
5 whole green onions, green and white parts diagonally sliced into
   2-inch lengths
2 tablespoons dry sherry

1. Place the chicken (including the neck if available) in a pot just large enough to hold it. Rub the chicken all over with the salt, including inside the cavity. Add the ginger, green onion, and sherry. Fill the pot with enough water to just cover the chicken by 1/2 inch.

3. Bring the water to a boil. Using a slotted spoon, skim away the brown scum that rises to the top and discard (it's harmless but unattractive). Partially cover the pot and reduce the heat. Simmer the chicken 15 minutes.

4. Turn off the heat. Leave the chicken fully covered in the pot for 1 hour. If desired, refrigerate overnight before use. Before serving, skim or remove the fat and discard the cooked ginger, garlic, and green onions. Serve the meat and broth as a soup, or separately (see headnotes).

## HAZELNUT CHICKEN SALAD ON SHREDDED NAPA CABBAGE
Serves 4

### Green Meter

- Green Goodness: No added fuel; hazelnuts and cabbage stretch cooked chicken
- Prep/Cooking Times: 10 minutes
- Prime Season: Year-round
- Conveniences: Can be partially made 1 day in advance and refrigerated
- New Green Basic: Use nuts and vegetables to stretch the meat; put leftover chicken to use.

*Ginger Chicken and Broth, Passively Poached (page 177) yields perfectly moist chicken for this distinctive salad, but any cooked chicken will do. Napa or Chinese cabbage is sweeter and milder than other cabbage and adds a crisp, frilly contrast to this dish. (To shred, stack the leaves and slice into 1/8-inch-wide strips.) Roasted hazelnuts bring crunchy sophistication, and nuts are nutritious, sustainable ways to cut back on carbon-intense proteins. Most domestic hazelnuts hail from Oregon, some from Minnesota. Most napa cabbage comes from California and Florida but may also be local to your area. It keeps for 2 weeks or more in the crisper—much longer than lettuce. As this recipe shows, it's tasty fresh and raw; as it ages, cook it into stir-fries or soups.*

3 cups shredded cooked chicken meat
1/4 cup rice vinegar
3 green onions
1/3 cup mayonnaise
3/4 cup coarsely chopped roasted hazelnuts, plus extra for garnish
8 ounces (about 3 cups packed, or 7 leaves) finely shredded napa
    cabbage leaves
1/2 teaspoon freshly ground pepper

1. Mix the chicken and vinegar in a bowl until the vinegar is absorbed. Trim and diagonally slice the green onions (both green and white parts) about 1/4-inch wide. Stir the mayonnaise and hazelnuts into the chicken. (The salad may be refrigerated for 1 day before continuing.)

2. Just before serving, mix in 3/4 of the cabbage, most of the green onion (reserve a bit for garnish), and pepper. Spread the remaining cabbage on a serving platter, spoon the salad on top and garnish with green onion and hazelnuts.

**Roasted Hazelnuts:** Roast the nuts whenever you're already baking something between 350 and 375 degrees F. Roast in a single layer on a baking sheet, 10 to 12 minutes, stirring once, or until the skin crackles all over. Pour into a clean towel and rub vigorously to flake off most of the skin (do this outdoors for less mess). Refrigerate or freeze, and toss into salads, pastas, breads, and desserts.

# SOUTHWEST SKINNY CHICKEN

*(A pressure cooker recipe)*

Serves 4

## Green Meter

- Green Goodness: Pressure cooker saves fuel; oven alternative, keeps kitchen cool
- Prep/Cooking Times: 12 minutes prep + 10 minutes marinating + 20 minutes cooking, mostly unattended
- Prime Season: Year-round
- Conveniences: Low-fat; morphable—serve as pieces or in soups, salads, sandwiches, or tortillas; repurpose excess broth into a soup or as a cooking liquid for beans or rice.
- New Green Basic: Use as a basic pressure cooker method for a whole chicken.

*Conventional roasting takes an hour, but in a pressure cooker, the bird is done in the time it takes to whip up a salad. I love the chile-lime flavor in summer, when pressure cooking makes such a logical alternative to a hot oven. Good winter seasonings are fennel seed, fennel bulb, and lemon; smoked or sweet paprika and tomatoes; or hoisin sauce, garlic, and ginger.*

*Instead of browning the skin, pull it off by hand (it takes just a minute) and discard. The bird still packs flavor, but it's very low-fat with no grease to pool in the copious juices. Marinate the bird for 10 minutes in the cooker, bring the pot to high pressure, and cook for 7 minutes. The result: tender chicken, with a luscious, spicy, brothy sauce. Easy, tasty, and greenly good.*

5 cloves garlic, chopped
Juice from 3 limes
2 teaspoons ground cumin
1 teaspoon dried marjoram
1/2 teaspoon dried thyme
2 teaspoons ground red chile (such as pasilla)
1 teaspoon salt
1 whole chicken (about 4 pounds), without giblets

2 carrots, sliced diagonally into 1-inch lengths
2 ribs celery, sliced diagonally into 1-inch lengths
1 onion, halved and thinly sliced
Warm flour tortillas (optional)

1. In a small bowl, combine the garlic, lime juice, cumin, marjoram, thyme, chile, and salt.

2. Skin the chicken by pulling the skin off with your hands, using a knife as needed. (Hold the skin with a paper towel to keep it from slipping.) Don't worry about the little bit of skin on the wings, which is almost impossible to remove easily. Discard the skin and any clumps of fat.

3. Place the chicken in the pressure cooker. Rub the spice mixture all over the chicken, pouring any excess into the pot. Marinate 10 minutes, or up to 1 hour.

4. Scatter the carrot, celery, and onion pieces around the chicken. Lock the lid in place and bring to high pressure. Cook, adjusting the flame as needed to maintain even pressure, for 7 minutes. Remove from the heat. Use the natural release method to let the pressure fall slowly. Serve the chicken in pieces with the vegetables and juices, and warm tortillas if desired.

## SPATCHCOCKED ROAST CHICKEN
Serves 4

### Green Meter

- Green Strategy: Flattening, convection cooking, and passive roasting halve both time and fuel
- Prep/Cooking Times: 10 minutes prep + 21 minutes active cooking + 15 minutes cooking unattended
- Prime Season: Year-round, but cold weather best for oven efficiency
- Conveniences: Easy to carve and serve
- New Green Basic: Refer to this recipe as a guide for quicker-cooking spatchcocking, instead of roasting whole poultry.

*A whole roast chicken typically takes more than an hour to roast. This recipe cuts fuel usage and time in half, with results so tender and juicy, you may never go back to conventional roasting again. It combines spatchcocking, passive roasting, and convection cooking, but even without convection mode, it's a speedy little number. I've kept the seasonings down to just olive oil, salt, and pepper, so you can focus on method. And frankly, such simplified seasoning really lets the flavor of the bird shine through. It's homey, satisfying comfort food.*

*Butterflying or spatchcocking refers to cutting out the backbone and flattening the breastbone, so the bird resembles an open book (or, as some readers have noted, a flat frog as seen from above). The bird cooks quicker and more evenly; this method works great on the grill, too. Another benefit: a roasted spatchcocked bird is so much easier to carve than a whole one. The time you spend removing the backbone before cooking delivers a definite pay-off.*

*You may need to fiddle a bit with the times to fit your particular oven and cookware. If you don't have a convection oven, actively cook the bird a little longer to get it to the 140 degree benchmark in step 3, before continuing with passive roasting.*

1 whole chicken (about 4 pounds), neck and giblets removed
1–2 tablespoons olive oil
Sea salt and freshly ground black pepper

1. Set out a low-sided pan with a rack, large enough to support the flattened bird. Rinse the chicken and pat it dry. To spatchcock it, place it with the breast down on a cutting board. Cut away the backbone with poultry shears or a knife (reserve the back for stock or another use). Place the bird breast-side up and press down on the breastbone with your palms, to crack it and flatten the bird. Snip the wishbone to break it (this allows the breast halves to relax and makes carving easier). Place the chicken on the rack in the pan, skin-side down, with the neck facing one long length of the pan.

2. Start heating the oven to 425 degrees F on convection setting, with an oven rack in the center position. Rub the bird's surface with olive oil and season generously with salt and pepper. Arrange the chicken with the wing tips tucked under the breasts. The legs/thighs will be loose and floppy: arrange them with most of the skin facing upward (not underneath the bird; they should look like a frog's silhouette), jiggling them around to fit within the confines of the pan.

3. When the oven is hot, slide in the pan, with legs to the rear of the oven. Roast 18 minutes. Remove the pan and using an instant-read meat thermometer, check the temperature in the thickest part of the thigh (avoiding the bone). If the internal temperature reads about 140 degrees F, return the bird to the oven (again with legs in back), and actively cook 3 more minutes. OR: if the internal temperature is closer to 130 degrees, cook 5–7 minutes before proceeding; if it is closer to 145 degrees or above, don't add additional active cooking time, and proceed to the next step.

4. Turn off the heat, keep the door closed, and let the bird passively roast 10 minutes. Remove from the oven and let rest 5 minutes before carving. The combination of passive roasting and resting will raise the internal temperature to 160–165 degrees (the juices will run clear when the chicken is adequately cooked). Cut the bird into serving pieces and serve.

## SIMPLE ROAST CHICKEN BREASTS
Makes 4 to 8 portions

### Green Meter

- Green Goodness: Roasts 4–8 portions in 30 minutes for standard oven; 20 minutes if convection roasted. Good for simultaneous cooking (with other 350 degree dishes).
- Prep/Cooking Times: 5 minutes prep + 20–30 minutes cooking
- Prime Season: Winter, or cool weather
- Conveniences: 4 ingredients, easy. Cook as main course or as morphable chicken meat.
- New Green Basic: Use the method for cooking chicken breasts in quantity, with less fuel; use as a template for convection cooking.

*You'll be surprised at the brilliant results of this ridiculously simple recipe. With so many seasonings available these days, it's easy to forget the sublime perfection of just salt and pepper. Used alone, they let the flavor of the food itself shine through. (And if you've got them, you can really flaunt your sexy sea salts and exotic peppercorns here.) But don't feel obligated to restrain yourself. This method works equally well with spice rubs.*

*Ovens aren't very efficient, but if you cook a lot of breasts, you'll make the most of the oven's heat, especially when cooking other dishes simultaneously. And if you need a large quantity of moist cooked chicken meat for salads, sandwiches, soups, or creative cookery, this pain-free, mess-free method yields tender, flavorful meat, perfect for absorbing other flavors. Depending on your oven's size, you can even roast two trays of 8 to 16 chicken breasts, swapping them midway as they cook. Just don't overcook the meat, or you'll lose its tender, juicy appeal.*

> 4 to 8 chicken breast halves, with skin and bone
> 1–2 tablespoons extra-virgin olive oil
> Salt and freshly ground pepper

1. Preheat the oven to 350 degrees F, at regular setting or in convection mode.

2. Arrange the chicken on a large baking sheet with low sides. Rub the pieces all over with olive oil. Generously salt and pepper all sides. For a standard oven, roast about 30 minutes. For convection roasting, cook about 20 minutes. (Chicken is done when firm but yields slightly to the touch, with clear juices; don't overcook; the hot pieces will continue carry-over cooking after they come out of the oven). Let rest 3–5 minutes before serving or refrigerate until ready to use.

# CHICKEN, APPLES, AND BRANDY, NORMANDY-STYLE
Serves 4

## Green Meter

- Green Strategy: Cooktop saves fuel and time (vs. oven); fruit stretches meat
- Prep/Cooking Times: 7 minutes prep + 25 minutes cooking
- Prime Season: Fall, winter, or when local apples are available
- Conveniences: quick 'n' easy; few ingredients; stretchable with noodles or bread
- New Green Basic: Use as a cooktop alternative to oven-braised chicken.

*This recipe has gone through several makeovers. I switched from oven-roasting to cooktop and actually prefer the results. Cooks in Normandy pour in cream, but to lighten the richness without losing intensity, I use cognac (a type of brandy) to lift the flavors of tart apples, earthy onions, and brown sugar. Then, for a more local complexion, I started splashing in domestic brandy from California instead of cognac from France. Très bon!*

*Green peppercorns (freeze-dried, not in brine) zip up the dish with mild pungency, but black peppercorns are fine. If you have it, use fresh thyme instead of dried, and garnish with a few sprigs before serving. Granny Smith apples bring the right brightness and crispness, but if a similar variety grows in your area, use it instead. Apples have a remarkably long storage life, so domestic apples may be found almost all year, not just in fall or winter. I serve this dish with buttered egg noodles or crusty bread, and sometimes a shredded red cabbage salad.*

3 tablespoons brandy
2 teaspoons brown sugar
2 Granny Smith apples (10 ounces)
1 onion (5 ounces)
3 pounds chicken thighs (6 to 8)
Coarsely ground sea or kosher salt
2 teaspoons olive oil
1 tablespoon coarsely ground dried green peppercorns (or black
    pepper)
1 teaspoon dried thyme

1. Pour the brandy and sugar into a mixing bowl. Peel, core, and dice the apples, tossing them into the bowl as you go; stir to coat, to prevent them from darkening. Dice the onion and stir into the apples.

2. Rinse the thighs and pat dry. Trim away excess flaps of skin and fat. Season the skin side generously with salt.

3. Heat the oil in a Dutch oven (or large heavy skillet) over medium-high heat. Add the chicken skin-side down. Season the flesh side generously with salt. When the skin side is nicely browned (5–7 minutes), flip the pieces over. Brown the underside, about 5 minutes.

MAIN W/MEAT

4. Push the chicken to the sides of the pan. Tilt the pan and spoon off most of the grease. Add the apple mixture, then rearrange the thighs skin-side up, with the apples nestled between and under them. Sprinkle green peppercorns and thyme over the thighs. Cover, reduce heat to medium-low, and cook about 15 minutes, until the apples soften and the chicken cooks through. The mixture will give off juices and settle into being a lovely sauce with bits of apples for texture. Serve the thighs with the apple mixture spooned over or under them.

## LEMON-TARRAGON TOASTER OVEN TILAPIA

Serves 4

### Green Meter

- Green Goodness: Toaster oven saves fuel, cooks 4 servings; sustainable from USA
- Prep/Cooking Times: 5 minutes prep + 6 minutes broiling
- Prime Season: Year-round
- Conveniences: 7 ingredients, quick 'n' easy
- New Green Basic: Use as a template for cooking fish fillets in a toaster oven.

*The Environmental Defense Fund rates tilapia from U.S. farms as an "eco-best" fish. The closed-tank systems allow for fewer escapees and less pollution. The fish eat a total or mostly vegetarian diet and bear few or no contaminants, so they're healthy to eat on a frequent basis. Latin American tilapia isn't quite as stellar environmentally, but it's an affordable alternative. However, the level of contaminants in Asian tilapia and weak farming practices put their tilapia on the list of fish to avoid.*

*Use the baking sheet that comes with your toaster oven for this recipe. If using a standard broiler, stretch your fuel by increasing the servings, and make the whole meal in the broiler, with garlic bread and Zucchini and Feta Strips (page 227), for instance.*

*Spicy Variation: If you live in the West and can get locally made Mexican cotija (co-TEE-yah) cheese, it makes a terrific substitute for*

*Grana Padano or Parmesan in this recipe. Substitute lime for lemon juice, and add chipotle chile instead of tarragon.*

1 tablespoon butter
4 tilapia fillets (1-1/2 to 2 pounds)
2 tablespoons fresh lemon juice
2/3 cup mayonnaise
1/3 cup grated Grana Padano cheese (or Parmesan or Romano)
2–3 tablespoons chopped fresh tarragon leaves
Dash of paprika

1. Preheat the broiler. Lightly grease a small baking sheet with some of the butter. Rinse the fish, pat dry, and arrange on the baking sheet in a single layer. Drizzle 2 teaspoons of the lemon juice over the fish and dot with the remaining butter.

2. Broil 5 to 6 minutes, or until the surface layer can be flaked with a fork but the interior is still undercooked. While the fish cooks, mix together the mayonnaise, cheese, tarragon, and remaining lemon juice in a small bowl. (Reserve some tarragon for garnish.)

3. Spread the mayonnaise mixture over each piece of fish. Sprinkle paprika on top. Broil until the sauce puffs up and browns, about 3 minutes. Garnish with reserved tarragon, and serve.

## TOASTED GARLIC TROUT WITH LIME

Serves 2

### Green Meter

- Green Goodness: Sustainable fish, often local; use of preblanched garlic saves fuel
- Prep/Cooking Times: 5 minutes prep + 10 minutes cooking
- Prime Season: Year-round
- Conveniences: quick 'n' easy; healthy, high in omega-3s
- New Green Basic: Use as a basic recipe for skillet trout; use the blanched, toasted garlic method in other skillet recipes. Substitute fillets for butterflied trout.

*Rainbow trout is often overlooked but has a wonderfully mild yet distinctive flavor. The Environmental Defense Fund rates farmed rainbow trout as an "eco-best" fish because of its minimal impact on the environment. In this recipe, garlic cloves, mellowed by blanching, toast in the pan for a sweet crunch that harmonizes well with trout and lime. To save fuel and water, blanch the garlic whenever you're boiling water for another use, up to 4 days earlier. If you increase the number of trout, you'll probably need two skillets.*

3 tablespoons fresh lime juice
2 rainbow trout, cleaned and butterflied (3/4 to 1 pound)
2 teaspoons ground cumin
Salt
2 tablespoons olive oil
8 large cloves blanched garlic, halved lengthwise (see Note)
Lime wedges, for serving

1. Pour the lime juice into a shallow dish. Dip the trout into the lime juice, coating on all sides. Sprinkle the inside (flesh part) with cumin and a generous pinch of salt. Close the fillets (as if closing a book) and sprinkle cumin and salt on the skin surfaces.

2. Heat the oil in a large skillet over medium-high heat. When the oil starts to ripple, gently place both trout into the pan, keeping them closed. Add the garlic cloves, pushing the garlic to the sides of the pan. Cook the trout about 5 minutes, flipping the pieces over when the skin is crisp and golden. Cook another 5 minutes or so, until the inside flesh is cooked through (it will flake when gently prodded). While the trout cooks, stir the garlic pieces occasionally, until the edges become slightly crisp and start to color. (If garlic browns too much it turns bitter, so remove it when golden, and set aside.)

3. Remove the trout from the pan. Spoon toasted garlic on top of each piece, and serve with lime wedges.

Note: *To blanch garlic, place whole unpeeled garlic cloves in boiled water and leave in for 2 minutes. Remove from the water and peel (the peels should slip off easily). Refrigerate up to 4 days.*

# MAPLE MUSTARD SALMON WITH CRISPY SKIN
Serves 4

## Green Meter

- Green Goodness: Cast-iron pan and passive cooking stretch fuel use; no food waste of salmon skin
- Prep/Cooking Times: 2 minutes prep + 10 minutes cooking
- Prime Season: Year-round
- Conveniences: Flexible portions; can scale up or down, no chopping, mostly pantry ingredients
- New Green Basic: Use to create crispy salmon skin, without an oven; use as a model for cast-iron cooking; use domestic mustard and maple syrup as versatile pantry ingredients.

*When I long for crispy, crunchy salmon skin (which is also so good for you), I turn to this technique. Gravlax and carpaccio are excellent no-fuel methods of preparing salmon, essentially curing the fish with salt or citrus and no heat. But without cooking, the skin turns flaccid and unappealing, so it's typically discarded—a waste of food and healthy omega-3s.*

*This method actively cooks salmon just long enough to crisp the skin, then passively finishes in the hot pan off the heat. It's a good opportunity to put your cast-iron skillet (or other heat-retaining cookware) to use. The famous chef Escoffier advised that salmon should be served as simply as possible, and here it's served with an uncooked sauce of just two main ingredients: real maple syrup and Dijon mustard. Coarse sea salt and snipped chives enhance the dish, or improvise with your own finishes; a balance of sweet-tart or sweet-salty complements the fish best.*

1 tablespoon vegetable or olive oil
4 (6-ounce) salmon fillets with skin, patted dry
2 tablespoons maple syrup
4 teaspoons Dijon mustard
Coarse sea salt or kosher salt
Freshly ground black pepper
1 teaspoon finely chopped chives (optional)

1. Heat a large cast-iron or heavy skillet over medium-high heat until very hot. Add the oil (just enough to coat the bottom of the pan) and give it a few seconds to heat up. Place the salmon in the oil, skin-side down. Reduce the heat to medium and cook until the skin is crisp, about 5 minutes.

2. Cover the pan but leave the lid slightly ajar for steam to escape, remove from the heat, and let rest 3–5 minutes, until the flesh is cooked through but still slightly rare. (For less rare, flip the pieces over and let the retained heat from the skillet cook the flesh side.)

3. While the salmon passively cooks, stir the maple syrup and mustard together until combined. Spoon or drizzle the sauce over the serving plate, and add the salmon, skin-side up. Garnish with sea salt, pepper, and chives, and serve.

## HOME-STYLE GLASS NOODLES
Serves 4 as a main course, 6 to 8 as a side dish

### Green Meter

- Green Goodness: No-cook noodles and quick frying conserve fuel
- Prep/Cooking Times: 15 minutes prep + 15 minutes cooking
- Prime Seasons: Year-round
- Conveniences: Flexible ingredients; meat-free adaptable
- New Green Basic: Use as a template for no-boil bean thread noodles instead of boiled pasta; add or subtract ingredients as desired.

*Cellophane or glass noodles, also called "bean thread" noodles, are made of mung beans. They come in neat little bundles and look similar to rice stick noodles, but with a shinier patina. They're commonly served in soups or eggrolls, but here they're the main event. Glass noodles don't have much flavor by themselves; instead, they readily absorb any sauce or soup broth, and they have a wonderful, slippery texture.*

*My mother used to make this, so for me it's home-style comfort food with added green benefits. It's a cool kitchen dish since all you do is soak the noodles in tap water until soft, and it makes a light noodle dish in*

*warm weather. The noodles are mixed with aromatics and fried pork or*
*beef, sesame seeds, soy sauce, and chicken broth. Ingredient amounts are*
*flexible (season to taste); vegetarians can adapt the basic noodle dish using*
*meatless ingredients. This dish goes well with the Chino-Latino Vegetables*
*on page 232, stir-fried vegetables, or a brightly seasoned salad.*

8 ounces bean thread noodles
2 teaspoons sesame oil
3 cloves garlic, minced
1 tablespoon minced ginger
1 pound ground pork or beef (or ground turkey)
4 tablespoons soy sauce (Kikkoman preferred)
3 tablespoons toasted sesame seeds
2 green onions, thinly sliced on the diagonal
About 2 cups low-sodium chicken broth
Ground black pepper
Garnish: About 1 cup coarsely chopped fresh cilantro (optional)

1. Soak the noodles in enough tap water (tepid or hot) to cover, until soft and pliable, 15 to 30 minutes. Snip with scissors several times. Drain well.

2. Heat the sesame oil in a large skillet or wok over medium-high heat. Stir in the garlic and ginger and cook until soft, about 1 minute. Add the ground meat and cook, breaking into crumbly bits, until mostly cooked. Reduce the heat to medium. Stir in 3 tablespoons soy sauce, the sesame seeds, and the green onion. Cook and stir a few minutes, until the soy sauce colors the meat.

3. Dump in the noodles, pour the chicken broth over them, and sprinkle on black pepper to taste. Stir to mix, and continue to cook, stirring occasionally. When most but not all of the liquid is absorbed, turn off the heat. Let the mixture rest 5 to 10 minutes. The noodles will continue to absorb the liquid.

4. Just before serving, mix in the remaining 1 tablespoon soy sauce and cilantro, if using. Serve warm or at room temperature, passing additional soy sauce on the side if desired.

# CHERRY-CABERNET PORK TENDERLOIN WITH TOASTED ALMONDS

Serves 4

## Green Meter

- Green Goodness: Nuts and dried fruit stretch protein, are refrigerator-free and pantry-ready; stovetop cooking and cutting meat in half saves fuel
- Prep/Cooking Times: 5 minutes prep + 20–25 minutes cooking
- Prime Season: Year-round
- Conveniences: quick 'n' easy, uses handy pantry items
- New Green Basic: Use a skillet, instead of the oven, for thick pieces of meat (cut in half and use a lid). Employ thin-slicing, nuts, and dried fruit as strategies for doubling standard meat servings.

*Want an instant way to stretch meat? Nuts and fruit work magic, and they may be available right in your own backyard. In this recipe, a pork tenderloin (normally meant to serve two persons) stretches into four servings with the addition of almonds and dried cherries (or use dried berries, figs, plums, or raisins). Serve thinly sliced on a platter with a pan-sauce of fruit and wine, and top with golden almonds. Fill out the meal with couscous or wild rice, and a spinach salad on the side or under this dish. Dried fruits and nuts add pantry-convenience, without cold-carriers to transport them. (Try this same basic recipe with chicken, too.)*

*Chefs often finish tenderloins in the oven to cook the centers through. It's not a very efficient use of a big oven. Skip the hot Humvee, add a lid, and pan-braise them instead. Cut long pork tenderloins in half, too; they cook more quickly in a skillet or on the grill.*

1/4 cup cabernet wine
1 tablespoon balsamic vinegar
2 teaspoons honey (preferably local)
1 tablespoon lemon-pepper seasoning salt
2 teaspoons fennel seed
1 pork tenderloin (1 to 1-1/4 pounds)

3 teaspoons olive oil
1/2 cup slivered almonds
1/2 cup dried cherries (or other dried fruit)

1. Combine the cabernet, balsamic vinegar, and honey. Combine the lemon-pepper and fennel seed.

2. Rinse the tenderloin and pat dry. Cut in half, to make 2 short, plump pieces. Rub with 1 teaspoon olive oil. Roll in the lemon-pepper mixture, coating all sides.

3. Heat a medium skillet over medium-high heat. When hot, add the almonds. Toast, shaking the pan occasionally, 3 to 5 minutes. As soon as the almonds turn golden, pour them onto a plate to cool. Wipe out the pan with a dry cloth.

4. Heat the remaining 2 teaspoons olive oil in the same pan, over medium-high heat. When the oil is hot, brown the pork on all sides, 5–7 minutes; if the pan seems dry as the pork cooks, add another teaspoon of oil.

5. Remove the pan from the heat while you add the cherries and carefully pour in the wine mixture (it will sizzle). Cover with a lid and return to the burner. Reduce the heat to medium-low and cook 10–12 minutes, turning the pork halfway through. When the pork is done, the pieces feel firm but the surface gives slightly when pressed; the inside should be slightly pink.

6. Remove the pork from the pan and let rest 5 minutes. Slice into medallions, 1/3–1/2 inch thick. Arrange on a platter, spoon on the cherry sauce, sprinkle on the almonds, and serve.

## MISO-BRINED PORK LOIN CHOPS
Serves 6 to 8; can freeze half after brining

### Green Meter

- Green Goodness: Skillet cooking saves fuel and time (vs. whole roasted); less packaging
- Prep/Cooking Times: 10 minutes active prep + unattended brining 1–2 hours + 10 minutes cooking
- Prime Season: Year-round
- Conveniences: 4 ingredients, quick 'n' easy; can brine and freeze
- New Green Basic: Use as an oven-free method for cooking pork loin (slice into chops); brine and freeze half to speed prep later.

*Whole boneless pork loins are economical, but roasting one consumes an hour's worth of oven fuel. For fast cooking without heating up the kitchen, slice into boneless chops, brine them (a tasty step, but optional), and cook in a skillet or grill pan. You'll consume less than 10 minutes of cooktop fuel. Plus, a whole loin skips the foam and plastic of packaged chops (buy the meat from the butcher counter directly, if you can). A 4-pound loin yields about 8 one-inch-thick chops.*

*Today's pork loins can be so lean that they lack flavor and easily overcook, becoming dry and tough. But there's a solution, literally: brine them in a mixture of water, salt, and sugar. With a brine, meat proteins swell into plump, juicy cells full of flavor and moisture, in one to two hours. And in this particular solution, umami-rich miso paste replaces salt, and iron-rich molasses takes the place of sugar. (Umami is known as the fifth flavor sense, and it creates an instant boost of delicious. Look for miso in Asian markets and some supermarkets.)*

*Freeze half the chops after brining if you like, then thaw overnight in the fridge and cook as below. Serve with rice and a sesame-cucumber salad, or with buttered noodles and steamed broccoli, or couscous and cranberry-mesclun salad.*

Brine:
1-1/2 cups water
1/4 cup shiro miso (white miso)
1 tablespoon molasses (unsulphured, blackstrap, preferably)

3–4 pounds boneless pork loin, sliced into chops, about 1 inch thick
1–2 tablespoons vegetable oil or olive oil

1. Combine water, miso, and molasses in a large measuring cup or bowl, stirring until the miso is completely dissolved. Submerge the chops in the brine (I fit them flat in one layer, in a 12 x 8-inch baking dish). Refrigerate in the brine for 2 hours, or 1 hour if you're short on time.

2. Remove the chops from the brine and blot dry. Heat 1 tablespoon oil in a large grill pan or skillet over medium-high heat. Add the chops and cook 3 to 4 minutes per side, turning once. (If necessary, use 2 skillets or cook in 2 batches, adding more oil as

necessary.) Chops are done when browned on the outside and the center is still pale pink (around 145 degrees internal temperature). Let rest 5 minutes before serving.

## FAIR TRADE VANILLA-ORANGE PORK

Serves 4

### Green Meter

- Green Goodness: Less than 10 minutes of cooktop fuel; opt for fair trade, organic vanilla and cinnamon, and local organic pork if available
- Prep/Cooking Times: 15 minutes start to finish
- Prime Season: Year-round
- Conveniences: Uses mostly pantry staples, quick 'n' easy, no chopping
- New Green Basic: Use as an oven-free method for cooking pork loin (without brining); use as a guide for seasoning with pantry-ready, local fruit preserves.

*Very tasty! Vanilla adds a smooth, homey undertone to this recipe, and orange and cinnamon magically blend into one integrated flavor. Marmalade brings preserved fruit into play as an easy sauce base, and if you can find locally made marmalade, you'll be even greener. Save packaging and money by slicing the chops from a whole boneless loin, and opt for organic pork that's Animal Welfare Certified, if you can find it. For an energy-efficient dinner, serve with rice prepared in a rice cooker and Crisp-Cooked Microwave Snow Peas (page 214).*

2 tablespoons orange marmalade
2 tablespoons cider vinegar
1 teaspoon finely grated fresh ginger
1/2 teaspoon vanilla extract
1/2 teaspoon coarse sea or kosher salt
1/2 teaspoon ground cinnamon
1 tablespoon sesame oil
4 boneless pork loin chops, about 3/4 inch thick (1-1/2 pounds)

1. Stir together the marmalade, vinegar, ginger, and vanilla.

2. Sprinkle half the salt and half the cinnamon on one side of the chops. Heat a large skillet over medium-high heat. Add the sesame oil. When the oil is hot, add the chops, seasoned side down. As the chops brown, season the other side with salt and cinnamon. Cook the first side 2–3 minutes until browned, then flip the chops, cover, and cook another 1–2 minutes.

3. Uncover, pour the sauce over the chops, turn off the heat, and let it sizzle in the hot pan to thicken slightly. (Chops are done when slightly pink in the center, and will continue to cook as they rest. Do not overcook or they will be dry.) Serve the chops with the pan's marmalade sauce drizzled over them.

# BLUE OVEN RARE ROAST BEEF

*(with optional Blue-Cooked Red Potatoes)*

Serves 8

## Green Meter

- Green Goodness: Uses 20 minutes of fuel, plus preheating
- Prep/Cooking Times: 5 minutes prep + 2 hours unattended active/passive roasting
- Prime Season: Year-round; but cold weather is ecologically best
- Conveniences: 4 ingredients, quick 'n' easy; can roast potatoes in same oven (see below)
- New Green Basic: Adapt the blue oven strategy to other whole roasts and vegetables: start at high heat for a limited period, then turn off the oven and leave the door closed for a long period, until done.

*You'll be amazed, and so will your guests, at how perfectly rare to medium-rare and juicy this dish turns out. It's ridiculously easy, even for holidays, but then that's part of what makes a blue oven so appealing: most "cooking" happens after you turn off the heat and walk away. For a side dish, the Blue-Cooked Red Potatoes can cook at the same time. Got leftovers? Make roast beef sandwiches au jus, or hash. Be sure to buy a* choice *grade of rump roast (or prime); lesser grades are better suited for braising. For a standing rib roast, cook with active heat for 5 minutes per pound, then turn off the heat and leave in the oven for 1-1/2 to 2 hours, door closed.*

1 (4-pound) rump roast, choice grade, at room temperature
3 cloves garlic
Extra-virgin olive oil
Kosher or coarse sea salt and freshly ground pepper

1. Heat the oven to 475 degrees F with a rack in the upper-third position.

2. Stud the roast with garlic: slice the garlic cloves into halves or large pieces. With a knife tip, stab a dozen or so holes all over the roast and punch in the garlic pieces. Rub the roast with olive oil, and generously season all over with salt and pepper. Place in a glass baking dish, fat side up.

3. Slide the roast into the oven, on the upper rack. Roast 15 minutes for a thin roast, or 20 minutes for a plumper roast. Turn off the heat. Leave the oven off with the door closed for 1-1/2 hours. Remove from the oven and let rest 10 minutes before slicing. The center will be rare, with medium rare closer to the ends, and the ends will be delectably browned and crisp. (If testing with a meat thermometer, the transition from rare to medium-rare occurs at 130 degrees F.)

**Blue-Cooked Red Potatoes:** Place a rack in the lower third of the oven. Quarter 4 large red potatoes (1-1/2 to 2 pounds) and combine in a baking dish with several cloves of garlic, salt, pepper, and just enough olive oil to coat. Seal with a lid or foil. Place in the oven, on the lower rack; cook simultaneously with the roast, and remove when the roast is done.

# Meat-Free
# Mains 'n' Grains

## FETA PESTO AND 2-MINUTE PASTA
Serves 3–4

### Green Meter

- Green Goodness: No-cook sauce and passive pasta save fuel and keep kitchen cooler; meat-free
- Prep/Cooking Times: 10 minutes prep + 15 minutes cooking, mostly unattended
- Prime Season: Year-round
- Conveniences: Quick 'n' easy; few ingredients
- New Green Basic: Use passive cooking to replace old full-fuel method for pasta; use the sauce as inspiration for other no-cook or meatless sauces.

*The "2-Minute" part of the pasta title refers to boiling time over live heat. Conventional recipes call for 8–10 minutes active boiling, but I prefer these results. Letting the pasta finish cooking passively, with heat turned off, uses less fuel and also prevents overcooking, ensuring al dente tenderness with proper toothiness. You can chop the pesto ingredients in a hand blender, food processor, by hand, or grind in a mortar and pestle.*

Feta Pesto (makes 1 cup):
2 cloves garlic
1 cup fresh flat-leaf parsley (1 ounce), tough lower stems removed
1/2 cup shelled walnut halves or pieces (2 ounces)
1 cup crumbled feta (2 ounces)
1/4 cup extra-virgin olive oil
1/2 teaspoon coarsely ground black pepper
2 teaspoons white or red wine vinegar

2 teaspoons salt
12 ounces dried spaghetti

1. If using a hand blender or small food processor, finely chop the garlic. Add the parsley and coarsely chop. Add the walnuts and chop until grainy. Add 1/2 cup of the feta and all of the oil. Process just until the mixture turns to a coarse paste. Pulse in the pepper just until combined. Set aside, and stir in the vinegar just before tossing with the pasta (to prevent discoloration).

2. Fill a large pot with about 5 quarts water and the salt. Cover and bring to a boil over medium-high heat. Stir in the spaghetti until completely submerged. Partially cover and cook 2 minutes, making sure the water doesn't boil over. Stir again to mix up the pasta. Cover, turn off the heat, and let stand until al dente, 8 minutes. (Remove the pot from an electric burner, so it doesn't boil over from residual heat.) Taste to see if the pasta is al dente; if not, allow another 1–2 minutes. Scoop out the pasta to drain.

3. When the pasta is mostly drained, toss with the pesto. If the mixture seems dry, mix in a spoonful of the pasta cooking water. Serve warm or at room temperature, passing remaining 1/2 cup feta on the side.

# MEATLESS MOUSSAKA
Serves 6 to 8

## Green Meter

- Green Goodness: Multiple methods cut traditional fuel in half; see all green strategies below
- Prep/Cooking Times: 25 minutes prep + 25 minutes cooking, mostly unattended

- Prime Season: Summer, fall for eggplant; substitute seasonal produce at other times of year
- Conveniences: Cook in advance and reheat; low mess
- New Green Basic: Use as a template for vegetable casseroles; replace bechamel or white sauce in conventional casseroles with this fuel-saving version.

*This is my poster child of fuel-efficient casseroles: it combines several New Green Basic strategies. It's also meatless: no cow carbon in the cookprint. The traditional Greek counterpart layers precooked eggplant and potatoes with tomato sauce and a classic white sauce, known as bechamel. Both sauces are made separately, so it's a pretty labor- and fuel-intensive dish.*

   *Green strategies start with the bechamel: you don't cook it separately; instead, a custard-like sauce bakes as part of the finished dish. Plus, this recipe shaves off at least 30 minutes of fuel from the traditional version, just in the time it takes to bake. Additionally:*

- *The vegetables precook in a microwave rather than on a cooktop*
- *The casserole can start in a cold oven to save fuel, or be added to a hot oven already in use (temperature and cooking times are flexible)*
- *It bakes passively for the last 5 minutes, fuel-free*
- *Two pie pans cook faster than one large baking dish (and leftovers take less fridge space). A single pie pan can cook in a toaster oven, if halving the recipe.*
- *It cooks faster in a large oven if placed in the upper third*
- *The recipe saves washing water through microwave cooking and less cookware (two pie pans and one mixing bowl replace a skillet, a pot, a colander, a saucepan, and various mixing bowls).*

*Use other cooked or grilled vegetables—summer squash, mushrooms, fennel, and/or bell peppers—or add some olives. If you don't have two glass pie pans (deep dish or regular 9-inch), other pans will work, like 8-inch-square cake pans or a single rectangular baking dish. Small eggplants can replace one large eggplant.*

1-1/2 pounds small russet potatoes (5 to 8), scrubbed
1 large eggplant (about 1 pound 6 ounces)

White Sauce:
1/2 cup heavy cream
1/2 cup plain yogurt
3 large eggs
1/4 teaspoon ground nutmeg
1/4 cup Pecorino Romano

1-1/2 teaspoons ground cinnamon
1 jar (24 to 26 ounces) prepared pasta sauce (Muir Glen Fire-
    Roasted or a marinara, preferred), or home-cooked sauce
Small amount of olive oil to grease pans
Pinch of salt
8 ounces crumbled feta (about 2 cups)
Freshly ground black pepper

1. Microwave the potatoes: poke 2–3 times with a fork and arrange in a glass pie
pan. Microwave on high, flipping them once, 5–7 minutes or until easily pierced.
Remove and let rest until cool enough to handle. Slice into 1/4 inch rounds (it's okay
if they fall apart or crumble).

2. Microwave the eggplant: poke with a fork about 12 times all over. Place in a glass
pie pan. Microwave on high 7 minutes, flipping once, until a fork inserted in the flesh
meets no resistance. The eggplant will slowly collapse after removal. Let the eggplant
rest until cool enough to handle. Cut into 1/2-inch-wide slices, which will be very soft.

3. Mix the white sauce: measure the cream and yogurt sequentially into a large
glass measuring cup (to save washing another bowl); add the eggs and nutmeg. Beat
with a fork until the eggs are mixed into the sauce. Beat in the Pecorino Romano.
Add the cinnamon to the jar of pasta sauce; stir, or reseal and shake to combine.

4. Assemble: wipe the pie pans clean, if needed, and coat with a small amount of
olive oil. For each pan, pour in 1/2 cup pasta sauce as the bottom layer. Arrange the
potatoes over the sauce (crumbled ones are fine, too). Season with salt. Sprinkle on
the feta. Layer on the eggplant. Spoon the remaining pasta sauce evenly over each
pan. Finish with a generous grind of pepper and an even layer of white sauce.

5. Bake in a cold oven: arrange a rack in the upper-third position. Place the pans
on the rack and set the temperature to 400 degrees F. Bake 20–25 minutes, or until

the sauce bubbles on the sides, and white sauce is set and lightly browned on top. Turn off the heat and let passively bake 5 minutes.

**Hot oven method:** bake in an oven heated to 350–400 degrees F for 20–30 minutes. Turn off the heat when the top starts to brown and let passively bake 5 to 10 minutes.

**Toaster oven method:** make half the recipe and layer the ingredients in a glass pie pan or 8-inch-square glass baking dish. Bake as above, in a cold or hot toaster oven.

# SHORT-CUT, PASSIVE LASAGNA
Serves 6

## Green Meter

- Green Goodness: Cooks with half the fuel, using regular noodles, cold oven, and passive baking; meat-free
- Prep/Cooking Times: 30 minutes prep + 45 minutes cooking, mostly unattended
- Prime Season: Cold weather months
- Conveniences: Less mess and cookware to wash; uses handy pantry ingredients
- New Green Basic: Modify conventional lasagna and pasta casseroles with this pasta preparation and reduced-fuel methods.

*Most lasagna recipes take at least an hour to bake; this recipe cooks with about 30 minutes of active fuel—and starts in a cold oven. I also take a noodle short-cut—but without using no-boil noodles, which are typically dried, then cooked and dehydrated at the factory. At home, they require extra sauce and/or cooking time to soften, and they often come out gummy. Plus, they cost more.*

*Instead, regular noodles with a passive hot soak work great. Fuel-wise, you still bring water to a boil, but you then turn off the heat and pour the water over the noodles to soak. For more energy efficiency, boil the water in an electric kettle and save the soaking water for your garden.*

*Pantry-ready olives, marinated artichokes, and roasted red peppers make winter meals snazzier and snappier. Same goes for pasta sauce in glass jars, which are greener than cans (though home-cooked sauce from fresh summer tomatoes is unbeatable, even if it's been frozen or home-canned). If possible, buy pantry goods in large containers, use what you need, then refrigerate the rest for other dishes. And recycle or repurpose the containers to cut down your cookprint. Cooked vegetables, like the microwaved eggplant on page 202, work super, too.*

9 lasagna noodles
3–4 teaspoons extra-virgin olive oil
15 ounces ricotta cheese (preferably whole-milk ricotta)
1 large egg
1 large clove garlic
1 teaspoon dried marjoram or oregano
1 (12-ounce) jar roasted red peppers, drained
1 (4.25-ounce) can chopped black olives, drained
1 (12-ounce) jar quartered marinated artichoke hearts, drained
1 cup (5–6 ounces) crumbled feta cheese
1 (24-ounce) jar prepared pasta sauce, or home-cooked sauce
8 ounces sliced or shredded provolone cheese (about 2 cups)

1. Bring about 6 cups water to a boil. Arrange the noodles in a 9 x 13-inch glass baking dish, in 3 stacks. Pour the water over them to cover, jiggling them around so they don't stick together. Cover with a baking sheet for best results, to help hold in the heat. Soak 25 to 30 minutes or until almost al dente (a bit longer soaking while you prep is okay). Separate the noodles once while soaking, so they don't stick together. Remove the noodles to a colander to drain. Drain the water from the dish, and wipe the dish dry. Oil the bottom and sides of the dish with a teaspoon or so of the olive oil.

2. While the noodles soak, beat the ricotta and egg together with a fork, until smooth and slightly fluffy. Pour the remaining olive oil into a microwave-safe mixing bowl. Mince the garlic and drop into the oil. Crumble the marjoram or oregano and stir into the oil. Microwave on high 30 seconds, to soften the garlic. Remove the bowl. Shred the peppers with fingers or chop them into large bites and add them to the bowl. Stir in the olives, artichoke hearts, and feta.

4. Start layering: spread a thin coat of pasta sauce in the bottom of the dish (about 1/4 of the jar). Top with a layer of three noodles. Spread all of the ricotta mixture over the noodles. Spoon on another layer of sauce (another 1/4 of the jar), gently pushing it over the entire ricotta surface. Add the second layer of noodles. Spread the pepper mixture evenly over the noodles. Add the final layer of noodles, spread on the remaining sauce, and top with a layer of provolone. Loosely cover with foil (oil the underside of the foil lightly to keep the cheese from sticking, if desired).

5. Place the lasagna in a cold oven. Turn the heat to 400 degrees F. Bake 25 minutes. Remove the foil, rotate the pan, and bake 5 more minutes. (The lasagna will not look done: the top cheese will be soft but not brown; sauce will show hints of bubbling around the edges.) Turn off the heat and let passively bake 15 minutes, or until the top browns and the sauce bubbles. (If the top still isn't brown enough, move to a top rack, flip on the broiler and broil for 1–2 minutes.) Remove from the oven, let rest 5–10 minutes to set, then slice and serve.

## SOBA NOODLES WITH TOASTED NUT OIL
Serves 4 as a side dish

### Green Meter

- Green Goodness: Passive cooking saves fuel
- Prep/Cooking Times: 5 minutes prep + <5 minutes cooking
- Prime Season: Year-round
- Conveniences: Quick 'n' easy; adaptable as a side or main; may be made in advance
- New Green Basic: Apply this pasta-cooking method (with or without tea) to other thin, delicate noodles; for thicker pasta, like spaghetti, cook up to 2 minutes before turning off the heat; use the no-cook sauce on other pastas and raw, spaghetti-sliced vegetables.

*Soba noodles are Japanese, but this version is only faintly Asian in flavor. Toasted hazelnut or walnut oil stands in for the usual sesame oil, but other flavorful oils work well, including olive oil, avocado oil, and basil oil. To serve the noodles as a tart salad pasta, brighten the mix with balsamic vinegar or Chinese black vinegar. White pepper isn't detectable at first, but it gives off its spicy heat slowly, reaching its full warmth at the end of a bite. Black or red pepper have their own nuances, also good.*

*Soba are made with buckwheat flour, sometimes flavored with green tea. For a home-cooked smoky tea flavor, cook the noodles in the same tea used for the Tea-Infused Eggplant Salad (page 228).*

*Soba noodles are perfect for passive cooking, which lessens the risk of overcooking them. Stir them into boiling water, turn off the heat, cover, and they'll be done in just over 2 minutes.*

6 cups water or brewed tea
2 bundles soba noodles (about 6-1/2 ounces)
2 tablespoons toasted hazelnut or walnut oil

## Quinoa: A Special Grain That's Especially Tasty Cooked the New Green Basics Way

I've never met a person who didn't like quinoa. It's nutty in taste and texture and is a powerhouse of nutrition. Quinoa lends itself well to salads, much like bulgur wheat does to tabbouli, or use it as you would rice (as a side dish or cooked into a one-dish main course).

The Incas, who considered quinoa a dietary staple, named it "the mother grain"—it's a complete protein, containing all the essential amino acids (making it perfect for meat-free diets). The small, round grains expand to four times their size when cooked. Most cookbooks say to boil and steam the grains like rice, or cook like pasta in boiling water, for about 15 to 20 minutes. The New Green Basics method starts the grains in cold water, gently boils them for 5 minutes, then lets them passively soften until done. Total time is comparable to, but uses less than half the fuel, of other methods.

**Rinsing:** Before cooking quinoa, rinse it thoroughly to remove the saponin, the natural insect-repelling coating that's bitter-tasting but harmless. Most quinoa today is prerinsed and needs just a quick splash at home, though bulk-bin quinoas may require more rinsing. To rinse, pour the quinoa into a fine mesh strainer and rinse under running

*(continues)*

3 tablespoons soy sauce
1/8 teaspoon ground white pepper
2 tablespoons balsamic vinegar (optional)

1. Bring the water or tea to a boil, covered, over medium-high heat. Add the noodles. Stir to loosen and separate. Cover, turn off the heat, and let the noodles passively cook until just tender, stirring once. Test after 2 minutes; they may need up to 4 minutes to reach al dente tenderness. Drain in a colander and rinse with cool water. Drain completely.

2. Toss the noodles with the oil. Add the soy sauce, white pepper, and vinegar, if using, and toss again to mix. Serve at room temperature. (Noodles may be kept 3 days refrigerated; bring to room temperature and refresh the seasonings before serving.)

(Quinoa: A Special Grain That's Especially Tasty
Cooked the New Green Basics Way, *continued*)

water. Or, place the strainer over a pot and run enough water over it to cover the quinoa. Swish the grains with your fingers to loosen the coating. Pour off the water and briefly hold the strainer under running water for a final rinse. How to tell when the saponin is gone? The water isn't soapy or bubbly, and a raw grain doesn't taste bitter. Drain and cook as below, or toast in a dry skillet, or with a smidge of oil or butter, before cooking. (I've been told captured saponin water makes a natural insect repellent in a spray bottle but haven't tried it yet.)

**Cooking:** Quinoa may be served hot or cold. To make 3 cups cooked quinoa, rinse 1 cup quinoa and place in a deep pot with 6 cups water and, if desired, 1 teaspoon salt. Cover and bring to a boil over medium-high heat, making sure the pot doesn't boil over. Reduce the heat and boil gently 5 minutes, covered or lid slightly ajar to prevent boil-over. Turn off the heat, cover completely, and let passively cook about 5 minutes, or just until the grains have no white dot in the center, have unfurled their spiral-like filaments (the germ), and still have a wee bit of crunch. Drain well; fluff the grains after 5 minutes. Serve hot or let cool and chill until ready to use (may be refrigerated up to 5 days).

# CURRY-SCENTED CARROT AND QUINOA SALAD
Serves 4–6

## Green Meter

- Green Goodness: Basic quinoa method saves fuel; meat-free
- Prep/Cooking Times: 10 minutes + quinoa cooking time (page 207)
- Prime Season: Year-round
- Conveniences: Quick 'n' easy; little chopping; versatile; do-ahead
- New Green Basic: Use as a nutritionally complete foundation for meatless entrees or hearty sides; substitute for pasta salads.

*You'll need cooked quinoa for this salad, and you can cook it up to 5 days in advance, whenever you're puttering around in the kitchen; or cook and mix directly into this recipe. Varying the flavorings is easy: omit the curry powder and skip heating the oil, and season as you would any grain salad. Cumin and chiles say olé; wine vinegar, basil, and feta croon the Mediterranean blues.*

*Curry powder tastes raw if not heated. To save washing another bowl, microwave the oil and spices in a glass mixing bowl, then mix in the rest of the ingredients.*

2 tablespoons extra-virgin olive oil
1-1/2 teaspoons mild curry powder
1/2 teaspoon ground cumin
3 tablespoons lime juice
2 teaspoons soy sauce
1 clove garlic, minced
1/4 teaspoon salt
1 large carrot
3 cups cooked quinoa (1 cup raw quinoa cooked as on page 207)
1/2 cup finely chopped red onion
1/2 cup chopped cilantro
1–2 serrano chiles, stemmed, seeded, and minced

1. Microwave the olive oil, curry powder, and cumin (stirred to mix) on high about 2 minutes, in 30-second bursts, or until the curry smells fragrant. (This takes away the rawness of the spices.) Stir in the lime juice, soy sauce, garlic, and salt.

2. Grate the carrot on the large holes of a grater. Pour the cooked quinoa and carrot into the bowl. Stir to combine. Stir in the red onion, cilantro, and as much of the chiles as desired. Serve at room temperature or slightly chilled.

# FRENCH LENTIL AND BULGUR "TABBOULI"

Serves 4

## Green Meter

- Green Goodness: One pot and passive cooking save fuel and water; meat-free
- Prep/Cooking Times: 40 minutes total, start to finish; mostly unattended
- Prime Season: Year-round
- Conveniences: One-pot meal; little chopping; adaptable
- New Green Basic: Use this method whenever you need cooked lentils or bulgur; apply combination one-pot cooking to other grain-legume recipes as a meat-free meal option.

*I consider this a starter recipe: it's here to get you started with technique and flavors, but after that, dive in and personalize it any way you like.*

*It's also basic to New Green cooking: the lentils boil for just 2 minutes in only 1 cup of water, then passively soak until done; at that point, the bulgur goes into the same pot, so it cooks in the same hot water (without added fuel). Although the lentils and bulgur cook together, if you deconstruct this recipe, you'll see that the same passive methods can apply to cooking the lentils and bulgur separately, if you need only one or the other.*

*Bulgur wheat may not instantly register in your culinary lexicon, but tabbouli may. It's at the heart of an iconic Middle Eastern salad, with tomatoes, herbs, and lemon. This recipe adds French green lentils, uses*

*more pantry-ready and year-round ingredients, and is good as a meatless*
*main course or as a side dish. Other variations? Nuts and dried fruit with*
*warm spices yield Moroccan flair; chiles and cumin can take the dish to*
*India with a touch of curry powder, or to Oaxaca with lime juice and*
*cilantro. You could toss feta and tomatoes into this mix, too, if you like.*

Lentils and Bulgur:
1 cup water
2 bay leaves
1/2 cup French green lentils, rinsed and drained
1/2 cup fine-grind bulgur wheat
1/2 teaspoon salt

Salad:
3 tablespoons extra-virgin olive oil
2 cloves garlic, minced
1/4 teaspoon crushed hot red pepper flakes, or to taste
3 tablespoons red wine vinegar
1 cup chopped Italian parsley
1 cup small diced red bell pepper
1/2 cup finely chopped red onion
1 tablespoon capers, chopped

1. Fill a small to medium-size pot with 1 cup water, the bay leaves, and the lentils. Cover and bring to a boil over medium-high heat. Reduce the heat and boil gently 2 minutes (make sure the pot doesn't boil over). Turn off the heat and let rest 15 to 20 minutes, or until the lentils are tender. (French green lentils hold their shape; they'll be firm, but not hard or crunchy, when tender.)

2. Stir in the bulgur and salt. Cover again and let rest 10 to 15 minutes, or until the bulgur is tender and the liquid absorbed.

3. While the lentils and bulgur rest, measure the olive oil into a mixing bowl. Stir in the garlic and red pepper flakes. Add the vinegar. When the lentil-bulgur mix is done, pour it into the mixing bowl and stir to coat with the dressing. Stir in the parsley, red bell pepper, red onion, and capers. Serve at room temperature now, or let the flavors mingle and serve lightly chilled. The salad will keep 2–3 days refrigerated.

# POLENTA WITH GOUDA
*(Standard oven and convection baked)*

Serves 4

## Green Meter

- Green Goodness: Bakes simultaneously with other oven dishes; finishes with passive cooking to save fuel; convection mode uses 1/3 less fuel than regular baking
- Prep/Cooking Times: 5 minutes prep + 35–50 minutes cooking, mostly unattended
- Prime Season: Winter, or when the oven's in use
- Conveniences: Mostly pantry ingredients, easy side dish
- New Green Basic: Use to replace cooktop-versions of polenta if the oven's already in use.

*Call it grits or polenta, cooked cornmeal makes for great eating and serious comfort food. Traditionalists cook it in a pot on the stove, with lots of stirring. Pressure cookers, slow cookers, and microwave ovens offer more fuel-efficient methods (discussed in Chapter 2 and at NewGreenBasics.com). But if the oven's already cooking, pop in this polenta dish to bake simultaneously. Without convection, allow 50 minutes of cooking, but convection mode cuts the total fuel time down to 35 minutes. Serve it as a hot side dish, or let it cool, and slice into squares the next day for grilling or pan-frying. The following instructions are written for both conventional and convection ovens.*

> 1 quart low-sodium chicken broth, or 1 quart water plus 1 teaspoon
>     salt
> 1/4 teaspoon white pepper
> 1 cup coarse-ground cornmeal
> 1 tablespoon extra-virgin olive oil
> 4 ounces shredded Gouda cheese (or cheddar), about 1 cup

**Convection oven method:** Turn on the convection mode to 350 degrees F. In a 1-1/2 to 2-quart casserole or baking dish, stir together the broth (or water and salt), pepper, cornmeal, and olive oil. Bake uncovered 30 minutes. Stir the cheese into the polenta. (Mixture looks soupy but comes together in the end.) Bake an additional 5 minutes. Turn off the heat and leave the polenta in the oven 10 minutes (if this is not possible, because other foods must continue cooking in the same oven, just leave the oven on). Remove it from the oven, stir, and let rest 5 minutes before serving.

**Conventional oven method:** Set the casserole on a rack in the upper third of the oven. Preheat the oven to 350 degrees F and bake 40 minutes, stir in the cheese, and bake 10 more minutes. Remove and let rest 5 minutes before serving.

# Vegetable Sides and Salads

## ASPARAGUS WITH LEMON, GARLIC, AND SOY
Serves 4

### Green Meter

- Green Goodness: Microwave saves fuel, uses no water, keeps kitchen cool
- Prep/Cooking Times: 5 minutes prep + ‹5 minutes cooking
- Prime Season: Spring
- Conveniences: Cook and serve in one dish; quick 'n' easy
- New Green Basic: Replace cooktop steaming of vegetables with this microwave cooking method; replace butter with this nondairy sauce on cooked or raw vegetables and pastas.

*Microwave ovens steam vegetables quicker than any other method, so green vegetables stay bright and tender. Some cooks snap off the woody ends of asparagus, but a vegetable peeler removes the tough skin without tossing out the tender center, so there's less waste and more good eating. The sauce here is similar to a Japanese ponzu sauce, but it's not really "Asian." It's just tasty, and versatile enough to complement flavors from France to Florida. Try it with broccoli, zucchini, mushrooms, and carrots.*

1 tablespoon fresh lemon juice
2 teaspoons soy sauce
1 teaspoon neutral vegetable oil, like canola oil
1 clove garlic, minced
1 pound fresh asparagus
Salt

1. Combine the lemon juice, soy sauce, oil, and garlic and let rest for 5 minutes for the flavors to marry.

2. Peel the tough skin from the lower ends of the asparagus (save the trimmings for stock or compost). Rinse the asparagus and leave some water clinging to the spears. Place in a microwave-safe baking dish and sprinkle with a touch of salt. Cook on high 1 minute. Turn the asparagus in the dish so they cook evenly. Continue to cook until crisply tender, 1–3 minutes, depending on the thickness and age of the spears. Pour any collected moisture out of the pan. Just before serving, pour the sauce over the asparagus, mix, and serve.

## CRISP-COOKED MICROWAVE SNOW PEAS
Serves 3–4 as a side dish

### Green Meter

- Green Goodness: Microwave saves fuel, uses no water, and keeps kitchen cool
- Prep/Cooking Times: <5 minutes prep + <3 minutes cooking
- Prime Seasons: Spring
- Conveniences: Can cook and serve in one dish
- New Green Basic: Replace cooktop vegetable steaming with microwaving; use for stand-alone side dishes or to parcook vegetables before adding to stir-fries or fried rice.

*Here's another microwave method that lets the fresh flavors shine and retains the crisply tender texture of delicate snow peas. It's a real time saver when you're making other dishes, and you can serve them straight from the cooking vessel, eliminating another pan to wash.*

1 teaspoon canola oil
2 cloves garlic, minced
1/2 pound snow peas
Salt to taste

1. Combine the oil and garlic in a microwave-safe bowl with lid. Cover and cook on high about 1 minute, stirring halfway through, until garlic is soft.

2. Rinse snow peas and drain. Leave some water clinging to the snow peas, then add the snow peas and salt to the bowl. Cover and cook on high 1 minute. Stir and test for doneness. Snow peas should be crisply tender; if too raw, continue to cook in short bursts until done, being careful not to overcook. Serve as a side dish or mix into a stir-fry.

# TENDER MICROWAVED CORN ON THE COB
Makes 2–3 ears of corn

## Green Meter

- Green Strategy: Microwave conserves fuel and water, keeps kitchen cool
- Prep/Cooking Times: 1 minute prep + 5–10 minutes cooking, partially unattended
- Prime Season: Summer, fall
- Conveniences: Quick 'n' easy; no big pan to wash
- New Green Basic: Replace boiling or cooktop steaming of corn with this microwave method.

*I've mentioned before that microwave ovens act as their own steamer vessels, trapping steam as effectively as a stovetop steamer, with little or no added water. Add this to the fact that they shave off considerable cooking time and produce few emissions, and you've got a friendly formula for low-impact cooking. If your microwave oven lacks a turntable, rotate the ears every now and then as they cook. Cooking times will vary depending on the microwave's wattage and the size and age of the corn, but it's a pretty idiot-proof method that yields tender results. For more ears, increase total cooking time by 2–3 minutes per ear.*

2–3 ears fresh corn on the cob, with husks
Salt, butter, and other seasonings as desired

1. Remove all but the last few (three or so) tender layers of husk. Cut off the loose ends and tassel. Rinse well and leave water clinging. Place on a microwave-safe plate; for 3 ears, arrange in the outline of a triangle for even cooking; or set the ears in a row and swap the inner and outer ears periodically during cooking.

2. Total cooking time is 5–6 minutes depending on size and number of ears. For even cooking: microwave on high 2-1/2 minutes. Spin the ears (so the top surface is now on the bottom; swap outer and inner ears at this time, too). Cook 2–3 minutes. Let steam in microwave 3–5 minutes. Use a towel to pull back the husks and remove the silks (the ears will be hot). Pull the husks all the way back over the stem, then cut them straight off to leave a fancy Elizabethan collar. Butter and season while warm, and serve on the same cooking plate. Variation: season with lime juice and chili powder, and salt to taste.

**Nonhusk method:** For naked ears, wrap in a double layer of damp kitchen towel. Be careful not to burn your hands from the steamy towels.

## "TEAPOT" OR PASSIVELY BLANCHED GREEN BEANS
Makes 2–4 servings

### Green Meter

- Green Strategy: Conserves fuel; captures water for reuse, and if using an electric kettle, kitchen stays cool; method applies to asparagus and other vegetables
- Prep/Cooking Times: 5 minutes prep + 5 minutes unattended cooking
- Prime Season: Summer, fall
- Conveniences: Quick 'n' easy cooking; versatile uses; keeps up to 4 days
- New Green Basic: Use to blanch vegetables instead of boiling them in a large pot of water.

*If you're a fan of crisply tender green beans for salads, you'll be amazed how well this method works, without much risk of overcooking. Boil the* water in an electric kettle, then pour it over the beans in a large baking dish. Use a conventional kettle (or covered pot) if you don't have an electric one.

Blanching slows the aging process and preserves freshness, and these beans will last most of the week if refrigerated. Besides being added to salads, they can be reheated in a microwave oven with butter, or added to pasta, soups, casseroles, and stir-fries. Make a big batch and use them in different ways throughout the week. You can use the same method with other vegetables, like sliced carrots and zucchini; just make sure the pieces are fairly small so the heat penetrates quickly.

1/2 to 1 pound green beans, trimmed
Salt (optional)

Cut green beans into 1 to 1-1/4 inch lengths. Place in a shallow glass baking dish and sprinkle with a few shakes of salt, if desired. Bring 5–6 cups water to a boil in an electric or conventional kettle, or enough water to cover the beans. Pour over the beans. Let the beans passively cook until just tender, about 5 minutes. Scoop into iced water or run under cool tap water (capture the water for another use). Drain and use as desired, or refrigerate up to 4 days.

**Asparagus:** Pencil-thin asparagus spears passively blanch in less than 5 minutes; cut thicker spears into shorter lengths. Trim and peel away any tough outer layer before blanching,

**Broccoli:** Broccoli is a dense vegetable, so it needs to be cut into small pieces. Remove the tough fibrous peel from the stalks. Dice the stalks into small, thin pieces (1/3 inch or less). Trim the floret stems into small pieces, too, and if the floret stems are thick, halve them vertically. Place a cover (like a pan or cutting board) over the broccoli after pouring in the water, to retain heat. Let passively blanch 6–7 minutes. This method works best if you want the broccoli slightly undercooked.

# ONE-POT PREP: POTATOES AND GREEN BEANS

Makes 1-1/2 pounds each of potatoes and green beans

## Green Meter

- Green Strategy: One pot and precooking save fuel and water; passive boiling saves fuel
- Prep/Cooking Times: 5 minutes prep + 5 minutes active cooking + 20 minutes unattended cooking
- Prime Season: Year-round for potatoes; green beans in summer and fall
- Conveniences: Shaves off cooking time later in other recipes; can be made in advance and keeps up to 3 days; less equipment to wash; adaptable to other vegetables
- New Green Basic: Plan ahead and cook multiple vegetables with about the same amount of water and fuel as one vegetable; use the basic passive method to replace continuous boiling.

*Use this method to save fuel, water, and your own time. It's a handy template for boiling potatoes and blanching vegetables together, even if you don't serve them in the same dish. Eat the vegetables separately or together, as in a Nicoise potato salad. Enjoy them right after cooking with a little butter and salt; or refrigerate and serve later in a salad, gratin, casserole, soup, Spanish tortilla, quiche, or omelet. Besides green beans, you can blanch a sequence of vegetables, assembly-line fashion, like carrots, broccoli, and asparagus, for example.*

*You'll need a large pot, a skimmer (like a Chinese spider or a large slotted spoon), a colander, and a bowl of chilled water (with ice or ice packs, or prechilled in the fridge) for shocking the beans. Shocking stops the cooking process and helps set the color. If the vegetables are cut in small pieces, though, cold tap water works almost as well as ice water. (Capture and repurpose the cooking and chilling water, for a greener cookprint.)*

*Smaller potatoes cook faster and don't get waterlogged like large, halved potatoes. Use any variety (Yukon Gold, red, russet, white). A wealth of nutrients lie just under the potato skin, so eat the peels, and put potatoes at the top of your organics list.*

1-1/2 pounds potatoes (2–3 inches in diameter, about 7)
1 tablespoon salt
1 to 1-1/2 pounds green beans

1. Scrub the potatoes but do not peel. Place in a large pot with the salt, and add water to cover by 2 inches. Cover the pot and bring to a boil over high heat. While the water heats, trim the stems off the green beans. Prepare a bowl of chilled or ice water.

2. When the water boils, add the green beans (they'll float above the potatoes) and push them down to submerge. Cook 2–3 minutes, until crisply tender or until desired degree of doneness.

3. Scoop the beans into the bowl of chilled water until cool. Drain in a colander. Leave the cold water in the bowl if blanching other vegetables, and add more ice or chill packs if needed.

4. The potatoes will not be done. Cover the pot, turn off the fuel, and let the potatoes passively cook 12–15 minutes, or until a skewer penetrates to the center. Scoop the potatoes into the colander to drain. Use now, or refrigerate the beans and potatoes separately (will keep 3 days).

## GREEN BEAN–WALNUT SALAD WITH FLAXSEED OIL DRESSING

Serves 2 to 3

### Green Meter

- Green Goodness: Fuel-free dish; stretches preblanched green beans
- Prep/Cooking Times: 5 minutes prep
- Prime Seasons: Summer and fall
- Conveniences: Can be doubled or tripled; quick 'n' easy; stretch into a main course (with tuna or tofu) or toasted cheese-bread on the side
- New Green Basic: Replace conventional green bean salads with this recipe. Use as a template for turning salads into hearty meatless meals, with nuts and substantial vegetables over leafy greens.

*Put your blanched green beans (from the One-Pot Prep on page 218) to use in this crunchy, flavorful salad. You can also use walnut, hazelnut, or olive oil, but flaxseed oil has a lovely nutty flavor all its own. Heating flaxseed oil destroys the nutrients, so salad dressings are a perfect way to enjoy its tasty benefits.*

*Flaxseed is especially high in omega-3 and omega-6 fatty acids (which help protect cell membranes throughout the body). The seeds and oil are packed with over-the-top nutritional benefits, from balancing hormones to lowering cholesterol (studies even show a spoonful a day can help combat dry eye). Flax is also the same plant that produces fibers for linen. Early Americans made clothing from it, and today, Minnesota and the Dakotas still produce almost all of the flax grown in this country. Even if you don't use flaxseed oil here, it's worth Googling for more info on its healthy properties.*

2 teaspoons flaxseed or walnut oil
2–3 teaspoons white balsamic vinegar
1 teaspoon coarse Dijon mustard, preferably Maille
1 clove garlic, minced
1/4 teaspoon salt
1/4 teaspoon freshly ground black pepper
2 tablespoons finely chopped red onion
6 ounces blanched green beans, in 1- or 2-inch lengths
1/2 cup (1.5 ounces) toasted walnuts, in large, coarse bits
2 tablespoons chopped fresh cilantro

In the bottom of a salad bowl, stir together the flaxseed or walnut oil, 2 teaspoons vinegar, mustard, garlic, salt, and pepper. Stir in the red onion. Add the green beans, walnuts, and cilantro. Toss to coat. Taste. You may need to add another teaspoon of vinegar and perhaps more salt to balance the flavors. Let the salad rest a few minutes for the flavors to marry, then serve.

# ROSEMARY-INFUSED POTATO SALAD
Serves 8

## Green Meter

- Green Strategy: Passive boiling saves fuel, cooks simultaneously in one pot of water
- Prep/Cooking Times: 40 minutes total combined prep/cook times
- Prime Season: Year-round
- Conveniences: Makes a large amount with little effort; adaptable to other flavorings
- New Green Basic: Replace conventional boiled potato salads with this more efficient cooktop method; use as a guide for cooking eggs and potatoes at the same time, to serve together or separately.

*Use this opportunity to cook extra eggs, for deviled eggs or snacks, or to passively blanch other vegetables at the same time, like green beans, carrots, garlic, and pearl onions (in their skins). Also, take advantage of the boiling water to skin tomatoes or peaches (cut an x in the base, then dip in boiling water for 10–20 seconds for the skins to slip right off).*

1-1/2 to 2 pounds small to medium russet potatoes, scrubbed
3 large eggs
3 sprigs fresh rosemary (optional)
1 tablespoon salt
2 ribs celery, with leaves

Dressing:
1/4 cup extra-virgin olive oil
Zest and juice from 1 whole lemon
3 tablespoons cider vinegar
2 tablespoons spicy brown or Dijon mustard
2 cloves garlic, minced
2 teaspoons finely chopped fresh rosemary leaves
1/2 teaspoon salt, or to taste
1/2 teaspoon freshly ground black pepper
3 tablespoons mayonnaise (optional)

1. Place the potatoes, eggs, rosemary sprigs, and salt in a large pot and add enough water to cover by 1 inch. Cover the pot and bring to a boil over high heat. Turn off the heat. Set a timer for 12 minutes.

2. While the potatoes passively cook, cut the celery into small dice and chop the leaves. Combine the dressing ingredients in a large mixing bowl. Fill a small bowl with ice water or chilled water to cool the eggs.

3. When the timer goes off, scoop the eggs out of the pot and into the ice water. Cover the pot again and let the potatoes passively cook until a skewer pierces the centers without resistance, 10 to 15 minutes. Scoop out the potatoes to cool slightly.

4. Shell the eggs and coarsely chop. Chop the potatoes into small, bite-size chunks.

5. Toss the warm potatoes in the dressing. Mix in the eggs and celery (including the leaves). Serve now or refrigerate (will last 3 days) and bring to room temperature; adjust salt and acidic flavors as needed before serving. If the salad seems dry, stir in the mayonnaise before serving.

## LEMON-MINT POTATO SALAD, SLOW-COOKER STYLE
Serve 6

### Green Meter

- Green Strategy: Slow-cooker method saves fuel, uses no water, and keeps kitchen cool
- Prep/Cooking Times: 15 minutes prep + about 2 hours unattended in slow cooker
- Prime Season: Summer, fall, spring
- Conveniences: Unattended cooking; less equipment to wash; flexible ingredients; can be made in advance and keeps up to 3 days
- New Green Basic: Replace cooktop boiled potato salads with this efficient no-water, slow-cooker method.

*If calories = heat, then this is one "low-cal" recipe: potatoes cook in a Crock Pot, without water, and the rest of the ingredients are garden fresh. Use this recipe as a template, or follow it to the letter for a breezy salad that's not heavy with mayonnaise but light with olive oil, herbs, and lemon.*

*After the potatoes cook unattended for about 2 hours in a slow cooker, they're ready to be mixed and seasoned any way you want. I use a 5-quart slow cooker, but larger and smaller ones work with some adjustments to cooking time. You'll need 1-1/2 to 2 lemons, and be sure to peel or grate the zest before juicing. Other fresh herbs to consider: basil, dill, cilantro, and marjoram. The dish makes a great base for tuna or a dressed green bean salad, too.*

> 1-3/4 to 2 pounds red potatoes (2–3 inches in diameter)
> 1 teaspoon salt
> 1/4 cup extra-virgin olive oil
> 1/4 cup fresh lemon juice
> Zest from the lemons (optional)
> 1/2 cup finely chopped red onion
> 3 tablespoons chopped fresh Italian parsley
> 2 tablespoons chopped fresh mint
> 3/4 teaspoon freshly ground black pepper

1. Scrub the potatoes and leave them wet. Poke each one with a fork twice, on opposite sides. Place in the crock of a slow cooker, cover, and cook on high 2 to 2-1/2 hours, or until just tender when poked with a skewer.

2. Slice into 1/4-inch-wide rounds while still hot (hold the hot potatoes with a fork while slicing, and don't worry about being exact; the potatoes will fall apart slightly when mixed). Toss with the salt first, then add the olive oil and toss. Pour in the lemon juice, and toss again. Then add the zest, red onion, parsley, mint, and pepper and toss until mixed. (Can be made earlier in the day. Cover, chill, and bring to room temperature before serving.) Taste to correct the seasonings; if too sharp or dry, splash in a bit of white balsamic vinegar or a little olive oil; or add a flavor-finish with something completely different, like basil oil or walnut oil. Leftover salad will keep refrigerated 3 days.

# FLUFFY GARLIC-RICOTTA POTATOES
Serves 6

## Green Meter

- Green Goodness: Passive cooking saves fuel; water can be repurposed; boiling water can be multitasked. Drop in extra potatoes and eggs for potato salad.
- Prep/Cooking Times: 5 minutes prep + 45 minutes cooking, mostly unattended
- Prime Season: Year-round, especially if multitasking and morphing
- Conveniences: Can be made 1 day in advance and reheated; morph leftovers into a browned gratin-like casserole
- New Green Basic: Use instead of conventional mashed potatoes; plan ahead and reserve half for baking as a different side dish when the oven's already in use.

*So simple, so comforting, so easy. Ricotta cheese and garlic make mashed potatoes tangy, creamy, and versatile: after mashing, the potatoes can be saved and reheated, without suffering in texture the way plain mashed potatoes do. If you're planning to oven-cook tomorrow night's entree (in a 350 to 450 degree range), simultaneously bake half the potatoes into the browned variation below. Don't peel potatoes (most of their nutrients lie just under the skin) unless you like them peeled, but do scrub them.*

*Potatoes for mashing should be very tender—unlike ones for potato salad, which profit from a touch of firmness—so actively simmer the potatoes in this recipe for a few minutes before you turn off the heat for passive cooking. Simmer larger potatoes for 10 minutes, smaller ones for 5 minutes. For extra flavor, cook the potatoes with fresh rosemary sprigs in the water.*

2 pounds Yukon Gold or russet potatoes
Salt
2 tablespoons butter
3 cloves garlic, minced
1/2 cup cream or half-and-half
15 ounces ricotta cheese (preferably whole-milk)

1. Place the potatoes and about 2 teaspoons salt in a medium pot and add enough water to cover by 1 inch. Cover and bring to a boil over high heat; reduce the heat and simmer about 10 minutes. Turn off the heat and let rest, covered, until potatoes are very tender when pierced with a skewer (20–30 minutes). Scoop out a half-cup or so of cooking water for later use. Drain the potatoes and water from the pot.

2. In the same pot, melt the butter over medium heat. Sauté the garlic until soft but not brown, about 2 minutes. Remove the pot from the heat and return the cooked potatoes to the pot. Break up with a potato masher, then gradually mash in the cream or half-and-half and ricotta. If too thick, thin with a bit of the cooking water. The potatoes should be light and fluffy. Add salt to taste, reheat if necessary, and serve. (To reheat, stir in some cooking water and warm over medium-low heat, or bake as below.)

**Golden-Brown Ricotta Potatoes:** Spread half the prepared potatoes into a greased baking dish. (Cover and refrigerate up to 2 days, if desired.) With a fork, make deep striations on the surface. Drizzle 2 teaspoons cream or half-and-half over the surface and dot with 2 teaspoons butter. Bake in the upper third of the oven at 400 degrees, or with other dishes between 350 and 450 degrees, until the center is warm and the top ridges browned, about 20 minutes. If the potatoes are warm but need more browning, briefly run them under the broiler.

# SUNNY SUMMERY SQUASHES

Serves 3–4

## Green Meter

- Green Strategy: Passive blanching with no additional cooking saves fuel.
- Prep/Cooking Times: 15 minutes prep + 10 minutes resting time
- Prime Season: Summer, fall
- Conveniences: May be made in advance; little equipment to wash; extra blanched squash can be morphed into other dishes
- New Green Basic: Replace conventional blanching with this method. Replace leafy salads with this more substantial dish to complement meatless mains.

*Bright green, orange, and deep red hues in this recipe virtually scream "Summer!" and "Fresh!" and so does the taste. Thinly sliced zucchini and yellow squash get a passive blanching treatment here, keeping them crisp and tender without tasting raw. Prepare extra and reserve for tossing into casseroles, omelets, salads, or stir-fries, or layering into a panini sandwich. Three garlic cloves sound like a lot, but passive blanching mellows their flavor and makes peeling easy.*

*Spoon the sun-dried tomatoes with some of their oil into the sauce as you measure them. (Some brands come with herbs in the oil, and make this an even tastier dish.) The squash and sauce components of this room-temperature dish can be prepared separately in advance, then mixed together at serving time (slice the basil and mix in right before serving, too).*

1/2 pound mixed summer squash (zucchini, yellow, pattypan)
3 cloves garlic, unpeeled
1/4 cup julienned or chopped sun-dried tomatoes in oil
1 tablespoon extra-virgin olive oil
1/2 teaspoon salt
1/4 teaspoon crushed red pepper flakes
1/2 cup shredded carrot
2 tablespoons grated Parmesan-Romano blend (optional)
2 tablespoons thinly sliced fresh basil

1. Bring about 4 cups of water to a boil (in an electric kettle or on the stove). While the water heats, slice the squash very thin, 1/8–1/4 inch thick, on the diagonal. Place in a shallow glass baking dish with the garlic. Pour in enough water to generously cover the squash. Passively blanch 2–4 minutes, stirring once or twice, or until just tender. Remove the squash to a colander to drain, and set the garlic aside. (Before using, if the squash are still wet, blot up moisture. Squash may be refrigerated 2 days but taste fresher if used the same day.)

2. Slip off the peels and finely chop the garlic. In a small bowl or the serving bowl, mix together the sun-dried tomatoes with some of their oil, olive oil, garlic, salt, and crushed red pepper. Let the flavors marry 5 minutes before using.

3. Toss the squash with the shredded carrots, oil mixture, and cheese, and let the flavors blend for 10 minutes. Mix in the basil, taste to adjust seasonings as needed, and serve.

# ZUCCHINI AND FETA STRIPS
Serves 4

## Green Meter

- Green Goodness: Piggybacks a hot broiler, conserves fuel; organic, local zucchini are easy to find
- Prep/Cooking Times: 5 minutes prep + 5 minutes cooking
- Prime Seasons: Summer and fall
- Conveniences: Broil before or after another dish; flexible amounts; 6 ingredients; quick 'n' easy
- New Green Basic: Use to replace other vegetable dishes whenever the broiler's hot.

*This is hands-down my favorite way to cook zucchini: tender strips, quick-broiled with feta cheese. For fuel efficiency, I plan to make this dish whenever I'm broiling something else. Select wide, plump zucchini to slice into thin strips. Italian seasoning is a robust mix of dried herbs, in the spice aisle, but oregano or marjoram work fine. Let the oil, herbs, and garlic marry for a while if you like, but you can also eyeball the amounts (which are flexible) without measuring or mixing: slice the zucchini, drizzle with olive oil, sprinkle on garlic and herbs, crumble on feta, and broil until browned. (It even tastes good cold.)*

2 tablespoons extra-virgin olive oil
2 cloves garlic, minced
1 teaspoon Italian seasoning, or dried oregano
2 or 3 large zucchini
4 ounces crumbled feta cheese
Freshly ground pepper

1. Preheat the broiler. Combine the olive oil, garlic, and Italian seasoning.

2. Slice the zucchini lengthwise, 1/8–1/4 inch thick. Arrange in a single layer on a large baking sheet. Brush or spoon on the olive oil mixture, distributing evenly. Drop nuggets of feta cheese liberally over the zucchini, and grind pepper on top.

3. Broil 3–5 inches from the heat, until the cheese browns, about 5 minutes. Let rest 3–5 minutes on the baking sheet (the zucchini will continue to cook from the pan's heat). Serve warm or at room temperature.

# TEA-INFUSED EGGPLANT SALAD

Serves 4

## Green Meter

- Green Goodness: Sub-boiling and passive cooking save fuel; tea may be reused to save water
- Prep/Cooking Times: 20 minutes, start to finish
- Prime Season: Summer, fall
- Conveniences: May be made in advance; quick 'n' easy; little chopping
- New Green Basic: Replace conventional eggplant salads with this recipe. Consider this cooktop recipe as an alternative to oven-baked or broiled eggplant. Apply the concept of steeping tea as a flavoring agent for other vegetables, pastas, shrimp, and poultry, instead of baking or roasting.

*Eggplant marries to smoky flavors, and steeping in Lapsang souchong tea brings intense smokiness to the palate without real smoke, in this riff on Korean eggplant salad. Some markets offer teas in bulk, so you can buy as much or little as you need. Repurpose the tea-infused water by using it to cook Soba Noodles (page 205), or in a brine for chicken, pork, or shrimp, or to fertilize plants (including the tea leaves). Two tea infusers, or two teabags, let the leaves unfurl and maximize the brew strength, but a single infuser will do.*

*Don't bring the water to a true boil here. By turning off the heat when the water temperature's still below the phase change (with some bubbles clinging to the pan but not circulating), you save fuel and time, the tea tastes better with more oxygen in the water, and the eggplant cooks more gently. It's okay if the pot jumps ahead to full boil; it won't hurt the dish, but it uses fuel unnecessarily.*

1 medium to large eggplant (1 to 1-1/4 pounds)
About 6 cups water
2 tablespoons Lapsang souchong tea (or 2 tea bags)

Dressing and Salad:
1 tablespoon toasted sesame oil

2 tablespoons rice vinegar
1 tablespoon soy sauce
2 teaspoons hoisin sauce
2 cloves garlic, minced
1 green onion, finely sliced on the diagonal
1/2 cup diced red bell pepper (corn-kernel size)
Salt and pepper to taste

1. Halve the eggplant lengthwise, then slice into wedges (like dill pickles). Cut into 1-inch chunks, each with some of the skin. (Chunks will seem large but shrink by half before serving.)

2. Fill a large pot with the water. Pack one or two infusers with tea (or use teabags) and drop into the water. Cover the pot, bring the water almost to a boil, and turn off the heat. Let tea infuse 5–10 minutes, stirring halfway through.

3. Remove the tea infusors or bags, cover the pot, and reheat to just below a boil. Stir in the eggplant (the tea will not cover the eggplant completely) and top with a pot lid. Turn off the heat. Steep the eggplant 5–7 minutes, stirring halfway through to mix the top layer and submerged chunks. The eggplant is done when softened and dusky in color. Scoop out the eggplant to drain in a colander in the sink (reserve the tea for other uses). When the eggplant is cool enough to handle, gently squeeze out excess water.

4. While the tea and eggplant steep, chop the salad vegetables and mix the dressing. Combine the sesame oil, rice vinegar, soy sauce, hoisin, and garlic. Mix the eggplant, bell pepper, green onion, and dressing, and season with salt and pepper to taste. Allow a few minutes for the flavors to marry. Serve at room temperature. The salad may be made 2 hours in advance; leftovers keep well up to 2 days, refrigerated.

VEGGIE SIDES

# AN-ROASTED MINI-TOMATOES
Serves 2; adaptable for more servings

## Green Meter

- Green Goodness: 5-minute cooktop alternative to oven baking; good use of local organics
- Prep/Cooking Times: 5 minutes prep + 3–5 minutes active cooking
- Prime Season: Summer and fall
- Conveniences: Flexible amounts; quick 'n' easy; 7 ingredients. Make extra for omelets, bruschetta, panini, or pizza.
- New Green Basic: Replace oven-baked tomato dishes with this cooktop method.

*I'll never forget the year my husband planted twenty-four tomato plants—just for the two of us. This recipe was one of the many, many, many tasty results: classically fresh and simple. The recipe can be doubled, tripled, or made with as many tomatoes as your pan can handle. For a crispy alternative to an oven gratin, serve the tomatoes sprinkled with fresh bread crumbs toasted in olive oil (toast the crumbs in the same pan and remove them before you add the tomatoes).*

1 tablespoon extra-virgin olive oil
1 garlic clove, minced
7 ounces cherry, teardrop, or other tiny tomatoes (about 1-1/2 cups)
2 teaspoons fresh lemon juice
Salt and freshly ground pepper
2–3 teaspoons chopped fresh mint or basil

1. Heat the oil and garlic in a medium skillet over medium-high heat. Roll the tomatoes around in the oil. Cook, shaking the pan often, until the tomatoes are warm and soft, 3 to 5 minutes, depending on size.

2. Add the lemon juice just before the tomatoes seem done, and shake the pan. Add salt and pepper to taste, stir in the herbs, and serve warm or at room temperature.

# BLUE OVEN CONVECTION-ROASTED SEMI-DRIED TOMATOES

Makes about 1 cup

## Green Meter

- Green Goodness: A fraction of the fuel and time used in traditional oven-drying; great for an abundance of organic, local tomatoes
- Prep/Cooking Times: 5 minutes prep + 20 minutes of active cooking + 2 hours passive drying
- Prime Season: Fall
- Conveniences: Flexible; can cook with other dishes in similar temperature range. Make extra for adding to main courses, casseroles, or pizza topping.
- New Green Basic: Replace slow-cooked oven-dried tomatoes with this alternative.

*As fall approaches and the weather cools, I consider oven-drying the season's overflow of tomatoes from our garden. But the thought of firing up the oven for hours, even at a low temperature, keeps me from doing it. This recipe is a good alternative. It convection-bakes the tomato halves on a rack for 20 minutes, then partially dries them in the closed oven for 2 hours. They're still a little soft and moist, which is good because they don't turn into rubber, and the flavors become delectably intense and concentrated.*

*For a greener cookprint, simultaneously cook these tomatoes with another dish, like a casserole, baked potatoes, or a gratin, and let them passively cook in the hot oven after the other dishes are removed, or together. You can also cook as many trays of tomatoes as your oven will hold to stretch your fuel expenditure.*

*I don't always recommend it, but in this case do line the pan with parchment paper or foil to catch the juices. The juices under the rack will dry out and blacken, and the pan will be a real challenge to clean if it's not lined.*

VEGGIE SIDES

10–12 Roma tomatoes
1 teaspoon Italian seasoning
Olive oil

1. Line a low-sided pan with parchment paper or foil, and set a rack on top. Stem, halve and seed the tomatoes. Arrange the tomatoes, cavity side up, on the rack. Sprinkle Italian seasoning and olive oil on top and in the cavities.

2. Arrange a rack in the upper third of the oven. Start the tomatoes in a cold oven, or heat the oven to 425 degrees F, preferably in convection mode.

3. Bake 20–25 minutes. With the door closed and heat off, leave the tomatoes in the oven for 2 to 2-1/2 hours. (It's okay to remove other dishes from the oven when they're done; just don't leave the door open for long.) Remove and use now, or store them in the refrigerator, coated in olive oil, for up to a week.

# CHINO-LATINO MIXED VEGETABLES

Serves 4 as a side dish

## Green Meter

- Green Goodness: Piggybacks a hot broiler to conserve fuel; may be stir-fried instead
- Prep/Cooking Times: 15 minutes prep + 8 minutes cooking
- Prime Seasons: Year-round, using seasonal vegetables
- Conveniences: Prep in advance; flexible ingredients; meat-free; quick 'n' easy
- New Green Basic: Think of this fuel-stretching dish when you plan to broil other foods; serve right away or later in the week, as a side dish or flatbread filling; use as a general recipe for mixed vegetables, broiled or stir-fried.

*If you're firing up the broiler for another dish, pop in these vegetables while it's plenty hot. Under high heat, the vegetables quickly roast, so cook as close to the broiler element as you safely can (raise the pan on an inverted baking sheet, for instance). They taste awfully good at room temperature, broiled before or after the main event. You can even cook while the broiler's hot, refrigerate, and serve tomorrow. As the title suggests, they're versatile enough to serve with Asian, Latin, or generic flavors. (For energy efficiency, don't fire up the broiler just for this dish; stir-fry the vegetables instead.)*

*Sometimes I add an (11-ounce) can of corn, rinsed and well drained; it stretches the mix to 6 servings, adds color, and transforms the mix to a meatless taco filling or topping. If you don't have green onions, half a yellow onion (thinly sliced) tastes fine, too; but it releases more liquid, which makes the mixture more apt to steam rather than brown.*

> 2 teaspoons sesame oil, plus extra for greasing the pan
> 1 tablespoon black bean–garlic sauce (I use Lee Kum Kee brand)
> 1 teaspoon Asian-style chile-garlic sauce
> 2 teaspoons sugar
> 2 large bell peppers (mixed colors, about 1 pound)
> 3 green onions
> 2 ribs celery, preferably with leaves
> 1 carrot

1. Lightly grease a large low-sided baking sheet with sesame oil. Combine 2 teaspoons sesame oil, black bean–garlic sauce, chile-garlic sauce, and sugar in a small bowl. Preheat the broiler until very hot (or prep the vegetables in advance, in step 2, and heat the broiler prior to cooking.)

2. Halve the peppers vertically and discard the stem, seeds, and membranes. Slicing from top to bottom, cut the peppers into thin strips, 1/4 to 1/3 inch at the widest point; line the strips up and diagonally slice into 1- to 1-1/2 inch lengths. Slice the green onions (green and white parts) on the diagonal into half-inch lengths. Before slicing the celery, remove the leaves and coarsely chop; reserve for garnish. Very thinly slice the celery and carrot, again on the diagonal.

3. Toss the vegetables in a large bowl with the sauce mixture just before cooking. Spread in a single layer on the oiled baking sheet. Broil close to the heat 4–5 minutes or until vegetables show signs of browning. Stir. Broil 3–5 minutes more, or until the vegetables are browned in spots and the tips are charred. Remove from the broiler and let the vegetables release steam, 5 minutes. Transfer to a serving dish, garnish with reserved celery leaves, and serve.

VEGGIE SIDES

# PARSNIP STRIPS IN COMINO BUTTER
Serves 4

## Green Meter

- Green Goodness: Fast, fuel-efficient, and water-free
- Prep/Cooking Times: 5 minutes prep + 5 minutes active cooking
- Prime Season: Fall, winter
- Conveniences: Flexible amount; quick 'n' easy; 4 ingredients; adaptable to other vegetables
- New Green Basic: Use instead of roasting, steaming, or boiling parsnips, carrots, and zucchini.

*A touch of butter and the warm scent of comino, or cumin, enhance the sweet, nutty flavor of parsnips. Like many root vegetables, parsnips can take some time to cook, but shaving them paper thin with a low-tech vegetable peeler speeds cooking and creates more surface area to sweeten and caramelize naturally. The parsnips can be prepared an hour in advance, too. Barely cook them, then turn off the heat. Right before serving, reheat until cooked through. Because the sugars in the parsnips slowly brown in the cooling and reheating process, they end up with even more richly browned edges.*

4–5 medium parsnips
2 tablespoons butter
1/2 teaspoon ground cumin
1/2 teaspoon salt

1. Using a vegetable peeler, peel the tough outer layer of the parsnips and discard (compost, or freeze with other trimmings for stock). Shred the parsnips into long, paper-thin strips. (If the core is woody, discard it with the scraps.)

2. Heat a large skillet over medium-high heat. Melt the butter, then add the cumin. Cook 1 minute. Stir in the parsnip strips and salt, coating well in the butter. Reduce the heat to medium. Cook 3–4 minutes, stirring occasionally, until the parsnips soften and become tender. Serve warm, or see the headnotes for reheating.

# ONE-STEP, ONE-DISH CAULIFLOWER GRATIN
Serves 3 to 4

## Green Meter

- Green Goodness: Cold start, toaster oven, and no separate sauce save fuel; one-dish cooking conserves washing water.
- Prep/Cooking Times: 12 minutes prep + 25 minutes baking, mostly unattended
- Prime Season: Winter
- Conveniences: One-pot dish; little chopping; few ingredients; can be doubled; can simultaneously bake with other dishes; flexible temperature range
- New Green Basic: Use to replace conventional baked cauliflower with cheese.

*This is inspired by a Bobby Flay recipe. I turned up the green dial by starting it in a cold oven and baking in a toaster oven. Flay's original recipe came with its own fuel-saving steps: (1) you don't need to blanch the cauliflower first; and (2) you don't need to make a separate cheese sauce before assembling the gratin. In fact, you don't even mess up a mixing bowl. It all goes into one ungreased pan, in layers.*

*This is such a flexible recipe. You may need to tweak the timing, but this dish works well in a small or large oven, with a hot or a cold start, and with or without other dishes. It may be doubled, and you can bake it in one baking dish or two pie pans. It can handle a range of 375 to 425 degrees, if it's baking with friends. It's an easy dish with results that are hard to mess up. And as flavor goes, what's not to love about cauliflower, cream, and toasted, melted cheese?*

1 pound (or slightly under) cauliflower florets (about 1 small head, or half a large head)
1/2 cup heavy cream
5 ounces soft goat cheese
4 ounces shredded Monterey Jack cheese
1/2 cup (3 ounces) grated Pecorino Romano (or Parmesan)
1/4 teaspoon salt
1/4 teaspoon freshly ground pepper

1. Layer the cauliflower into an 8x8-inch glass baking dish, or equivalent (like a deep-dish glass pie pan). Pour the cream over the cauliflower, crumble on the goat cheese, sprinkle on the Monterey Jack, and top with the Pecorino Romano. Season with salt and pepper.

2. Place the dish in a cold toaster oven. Set the oven to 400 degrees F and set a timer for 25 minutes. Check the gratin: the sauce should bubble visibly through the sides of the pan, and the cauliflower should be almost tender when pierced with a fork. If not quite ready, return the dish to the oven, turn off the heat, and let passively bake for about 5 minutes. When the gratin is toasty on top, let it rest 5 minutes out of the oven for the sauce to set. Serve hot.

# MAPLE-BUTTER ACORN SQUASH, SLOW-COOKED
Serves 2

### Green Meter

- Green Goodness: Slow cooker saves fuel; maple syrup is a natural, domestic ingredient
- Prep/Cooking Times: 5 minutes prep + about 2 hours unattended slow cooking
- Prime Season: Fall, winter
- Conveniences: Few ingredients; easy
- New Green Basic: Substitute this slow-cooker method for oven-baked squash.

*Winter squash take about an hour to oven-bake, even when halved. But a slow cooker consumes a fraction of the energy, even when compared to a gas oven, and gets tender, tasty results. (If you do bake squash, start them in a cold oven and turn the heat off early to passively finish cooking; or bake simultaneously with other dishes.)*

*You'll need a large round or oval slow cooker (5 quarts or so) for this recipe. Dedicated slow-cooker cookbooks detail all sorts of timing and techniques for sliced pieces, whole squash, and multiple squash, but this recipe captures the basics. Keep the water down to 2–3 tablespoons to prevent soggy squash, and cook entirely on high for faster results; or cook on low for 3–6 hours, depending on the squash and the slow cooker. To help a lopsided half sit evenly in the crock, without spilling its buttery juices, trim a thin piece from the bottom or the side and shim it up. Cut the squash into quarters or wedges if the halves don't fit easily in the crock, and reduce cooking time by 10–20 minutes.*

*Play with flavors. Other good squash partners are maple, chile, and lime; Chinese five-spice powder, brown sugar, and sesame oil; or garlic and browned butter, for instance.*

3 tablespoons water
1 acorn squash (1-3/4 to 2 pounds), washed
4 teaspoons butter
2 tablespoons maple syrup
1/4 teaspoon salt

1. Pour the water into the slow-cooker crock. Halve the squash vertically and scoop out the seeds and strings. Set the halves, skin side down, adjacent to each other in the crock. Evenly divide the butter, maple syrup, and salt into the cup of each squash.

2. Cover and cook on high about 1-1/2 hours, or until tender (but allow an extra 30 minutes on high depending on the squash and the cooker). Keep warm in the slow cooker, with heat off, until ready to serve.

# SQUEEZED RED CABBAGE SALAD WITH MINT
Serves 4

## Green Meter

- Green Strategy: Tenderizes without cooking; works with zucchini, carrots, and other vegetables
- Prep/Cooking Times: 10 minutes active prep + 30 minutes unattended soaking
- Prime Season: Fall, winter
- Conveniences: Quick 'n' easy cooking; versatile uses; can be made in advance; adaptable to other flavors
- New Green Basic: Use as a no-cook tenderizing method, instead of steaming or stir-frying.

*A whole salad of raw red cabbage can give your teeth and jaws a workout. Soaking shreds in a salty brine breaks down the tough, dense fibers, but in a marvel of food science, still keeps a tenderly crisp texture, which doesn't get lost even after 3 days in the fridge. Try this treatment in winter, when red cabbage is in season and tender greens are not. The cabbage works fine with any dressing. Lime and red chile spin Mexican, caraway and wine vinegar go German, and rice vinegar and sesame oil sing with Asian harmonies.*

2 cups water
2 tablespoons salt
1/2 small head red cabbage (12–14 ounces)
1 green onion
2 teaspoons honey
4 teaspoons apple cider vinegar, plus more if needed
3 teaspoons mild olive oil
1/2 teaspoon dried mint
Freshly ground pepper to taste

1. Stir the water and salt in a mixing bowl until the salt dissolves. Slice the cabbage into thin shreds (about 1/8 inch wide). Submerge the shreds in the brine. Soak, stirring once or twice if convenient, about 30 minutes, or until the white parts of the cabbage turn dusky. Drain in a colander, rinse, and squeeze the cabbage by the fistful to release liquid. As you go, drop the squeezed cabbage into a mixing or salad bowl.

2. While the cabbage soaks, thinly slice the green onion (green and white parts). Combine the honey, cider vinegar, olive oil, and mint. Toss the squeezed cabbage with the dressing, green onion, and pepper. If not serving immediately, chill 15 minutes or up to 3 days. Taste and add more cider vinegar if the flavors need refreshing. Serve lightly chilled or at room temperature.

# Breads and Sweets

## TRUE SKILLET CORNBREAD
Serves 4 to 6

### Green Meter

- Green Goodness: No-oven cooking saves fuel and keeps kitchen cooler
- Prep/Cooking Times: 5 minutes prep + 20 minutes cooking, mostly unattended
- Prime Season: Year-round
- Conveniences: Mostly pantry ingredients; quick 'n' easy side dish; can use an induction burner
- New Green Basic: Replace oven-baking with this cooktop cornbread method.

*Most "skillet cornbread" recipes are misleading: they bake the batter in a cast-iron skillet in the oven. This all-stovetop recipe is the real deal, no cheating, and no need to turn on the oven just to enjoy hot cornbread. Sure, the top of the loaf doesn't take on a luscious golden brown color, but the bottom does, and that's how you serve it: bottom up. (It's like making a thick corn pancake: golden and crunchy on the outside, tender in the*

*center.) You'll need a lid for this dish, and a wide spatula and plate for*
*flipping the loaf; if you've got an induction burner, use it here. (You'll also*
*find energy-friendly cornbread recipes in slow-cooker cookbooks.)*

3/4 cup cornmeal
3/4 cup unbleached all-purpose flour
1 tablespoon sugar
1 teaspoon salt
3/4 teaspoon baking soda
3/4 teaspoon black pepper
3/4 cup buttermilk (regular or low-fat)
3 tablespoons vegetable oil
2 large eggs

1. In a mixing bowl, stir together the cornmeal, flour, sugar, salt, baking soda, and pepper. Measure the buttermilk into a 2-cup or larger measuring cup, add 2 tablespoons of oil, and beat in the eggs just until blended.

2. Start heating a 12-inch cast-iron skillet over high heat. When the pan is almost hot, pour the liquids into the mixing bowl; mix with a few quick strokes, just until the wet and dry ingredients are combined. When the pan is hot, add the remaining 1 tablespoon oil, swirl it in the pan, and heat until the oil shows faint ripples (this happens quickly if the pan is already hot). Pour the batter into the pan. Cover, reduce the heat to medium-low, and cook 15 minutes (the edges will be brown and the top dry).

3. Slip a large spatula under the bread, and lift the loaf enough to slide a plate under it. Flip the plate over, using the spatula to hold the loaf in place, and drop the loaf back into the pan, browned side up. Turn off the heat and let passively cook, uncovered, 5 minutes. Slide the loaf out of the pan, slice into wedges, and serve hot, slathered with butter if desired.

# GREEK CITRUS-HONEY CAKE
Serves 12

## Green Meter

- Green Goodness: Bakes in slow cooker using almost no fuel; kitchen stays cool
- Prep/Cooking Times: 15 minutes prep + 2-1/4 hours unattended cooking
- Prime Season: Year-round
- Conveniences: Quick 'n' easy dessert; small slices serve many
- New Green Basic: Use as no-oven, slow cooker template for other quickbreads and cakes.

*This rustic and distinctive cake can be addictive with tea at breakfast, with cheese at lunch, and after dinner with grapes and fresh fruit. It's inspired by a cake in Lynn Alley's book* The Gourmet Slow Cooker, *and it's like the sweets served at Mediterranean cafés and coffee houses—moist with lemon-honey syrup, fruity with olive oil and oranges, spiked with cinnamon and yogurt. Cornmeal and almonds give it texture you can taste. And unless you mention it, no one will guess it's made in a Crock-Pot.*

*For easy mixing and one less bowl to wash, measure the wet ingredients starting with the yogurt into a 4-cup measuring cup, then add the eggs and whisk.*

1/2 cup olive oil (mild, or mix of mild and fruity), plus extra for
   greasing
1-3/4 cups sugar
1-1/2 cups unbleached all-purpose flour
1/2 cup cornmeal
1 teaspoon baking soda
1 teaspoon baking powder (double-acting)
1/2 teaspoon ground cinnamon
1/2 teaspoon salt
8 ounces (1 cup) plain yogurt (regular or low-fat)
2 teaspoons orange oil or 1 tablespoon orange extract
6 large eggs
1/2 cup slivered almonds, or pine nuts

**Syrup:**
1/4 cup honey
3 tablespoons fresh lemon juice

1. Grease the bottom and sides of a 5-quart slow-cooker insert (crock) with a small amount of olive oil. Cut a piece of parchment paper to fit the bottom. Set in the paper and grease it.

2. In a mixing bowl, stir together the sugar, flour, cornmeal, baking soda, baking powder, salt, and cinnamon. Separately combine the yogurt, olive oil, orange oil or extract, and eggs, beating with a wire whisk. Pour the yogurt mixture into the bowl holding the dry mixture and combine until uniformly mixed. Stir in the nuts. Pour the batter into the crockery insert.

3. Lay a folded dishtowel across the top of the crock (covering the batter without touching it), cover with the lid, and cook on high 2 hours and 15 minutes, or until the edges turn brown and pull away slightly from the insert, and a wooden skewer poked in the center comes out clean. Lift the insert (using potholders) out of the cooker and let it rest, uncovered, 15 minutes. Loosen the sides of the cake with a knife or spatula. Place a plate over the top and, holding it securely (it's hot: use potholders), flip the crock over, so the cake falls onto the plate. Remove the parchment. Let the cake cool slightly.

4. Stir the honey and lemon juice together until completely combined. While the cake is warm, poke holes in the top with a fork, about 20 times. Spoon the glaze over the top and sides, letting the glaze seep in slowly before adding more. Serve in thin slices.

# RICH CHOCOLATE MINI-MOUSSE WITH SEA SALT
Makes 8 servings

## Green Meter

- Green Goodness: Microwave prep and no-baking save fuel, keep kitchen cool; ideal for organic, fair trade chocolate
- Prep/Cooking Times: 20 minutes, start to finish; plus chilling time
- Prime Seasons: Year-round
- Conveniences: Quick 'n' easy; few ingredients; uses dairy but egg-free
- New Green Basic: Replace cooked desserts with this nonconventional, nearly fuel-free method; use the microwave technique for melting chocolate in other recipes.

*A serious, intense blast of chocolate can be exhilarating, as in this mini-dessert. Serve in very small portions, like sake cups or shot glasses. This is a basic, flexible recipe. Make it with regular cream instead of heavy whipping cream. Go high end with 70 percent or better organic specialty chocolate and gray sea salt, or use supermarket chocolate chips and regular sea salt. Beat the cream with a whisk, by hand, the whisk attachment of a hand blender, or with electric beaters. If chopping the chocolate by hand, shoot for easy-to-melt chip-size pieces. Go wild with other flavors, like cinnamon, mint, or orange liqueur, too; a few shakes of ground red chile make it really interesting. Got cacao nibs on hand? Toss 'em in or on for crunch.*

1 cup plus 1/4 cup heavy whipping cream
6 ounces chopped bittersweet chocolate or semisweet chocolate
  chips
1 teaspoon vanilla extract
2 teaspoons coffee liqueur or coffee syrup
1/8 teaspoon salt (preferably finely ground sea salt)
Coarse ground sea salt for garnish (optional)

1. Beat 1 cup whipping cream until thick enough to hold a peak.

2. Microwave the chocolate in a microwave safe bowl, on high, about 1-1/2 minutes, stirring 2 or 3 times, until melted. (Chocolate may not look melted but when stirred will become smooth and silky.)

3. In a microwave-safe measuring cup, heat the remaining 1/4 cup cream until warm, about 30 seconds, but don't let it boil. Stir in the vanilla extract, coffee liqueur, and the fine salt. Stir the warm cream mixture into the melted chocolate gradually, until smooth and combined.

4. Pour the chocolate mixture into the bowl of whipped cream. With a spatula, gently fold the chocolate into the whipped cream, taking care to keep the mixture as light and fluffy as possible (it's okay if some white or lighter colored streaks remain).

5. Spoon into 8 sake cups or shot glasses and garnish with grains of coarse sea salt. Refrigerate until set, about 2 hours; serve slightly chilled. (May be made 2 days in advance.)

# COLD-OVEN CLOVE AND CRYSTALLIZED GINGER CAKE
Serves 10–12

### Green Meter

- Green Goodness: Starts in a cold oven, finishes in a passive oven; saves fuel
- Prep/Cooking Times: 15 minutes prep, rising + 25 minutes active and passive baking
- Prime Season: Year-round
- Conveniences: Quick to mix, easy dessert; mostly pantry ingredients
- New Green Basic: Use as a template for baking without preheating and finishing baked goods under passive heat.

*Crystallized or candied ginger keeps indefinitely, and I was pleased to find it in the bulk bin of a local supermarket. The punch of ginger and cloves makes this semisweet yeast cake a good accompaniment to plums or pineapple with goat cheese or yogurt—or serve it as a coffee cake. Coarsely chop the ginger chunks into corn-kernel-size pieces.*

Cake:

3/4 cup water

4 tablespoons butter in chunks, plus some for greasing the pan

2-1/2 cups unbleached all-purpose flour

1/2 cup sugar

2 (1/4 ounce) packages dry instant yeast (4-1/2 teaspoons total)

1 teaspoon ground cloves

1/2 teaspoon salt

2 large eggs, beaten

Streusel:

1/4 cup oats (old-fashioned or quick cooking)

1/4 cup chopped crystallized ginger

1/4 cup packed brown sugar

2 tablespoons cold butter, in small chunks

1. Butter an 8-inch-square glass baking dish. Microwave the water and 4 tablespoons butter together on high 1 minute; stir to mix butter chunks into the hot water. If the butter needs more melting, cook 30 more seconds and stir. (Watch to make sure mixture doesn't boil over.) Mixture should register 120–130 degrees on an instant-read thermometer. (If hotter, let it cool so it doesn't kill the instant yeast.)

2. In a large mixing bowl, stir together the flour, sugar, yeast, cloves, and salt until well mixed. Pour in the water-butter mixture and stir to combine. Stir in the eggs until the mixture thickens into a uniform but wet dough.

3. Pour the dough into the pan. Cover with a kitchen towel while it rises. Set the timer for 25 minutes (to measure rising time).

4. Combine the oats, ginger, brown sugar, and butter, loosely rubbing the butter in with your fingers to mix (mixture should be rough and crumbly, not uniform). Top the dough with the oat mixture. Place the pan in a cold oven while it continues to rise (without the towel). When the 25 minutes are up, set the heat to 350 degrees F and reset the timer to 25 minutes. When the timer goes off, turn off the heat (the sides will be lightly browned). Let the cake passively cook 5 minutes. The sides and top will darken a bit more, and the interior will finish cooking. Remove the pan from the oven and cool on a rack (the cake will continue to cook as it cools). Serve plain, or with fresh fruit and yogurt.

# Index

Ginger Chicken and Broth, Passively
    Poached, 177–178
glass
    bakeware, 105–107
    food packaging, 142
    food storage, 150
    glass noodlesbean thread noodles
    Home-Style Glass Noodles, 190–191
    soaking, 4–5, 131
    Vietnamese Glass Noodles Soup with
        Chicken, 173–174
global warming
    food transportation and, 112
    impact on food chain, 11–13
goat cheese, in One Step, One Dish Cauliflower
    Gratin, 235–236
Golden-Brown Ricotta Potatoes, 225
Gorgonzola sauce, no-cook, 132
Gouda, Polenta with, 211–212
grains
    bulgur wheat, 5, 78, 130–131
    Curry-Scented Carrot and Quinoa Salad,
        208–209
    dominance in meal, 138–139
    energy-synergy tips, 4, 133–134
    French Lentil and Bulgur "Tabbouli," 209–210
    Polenta with Gouda, 211–212
    pressure cooking, 134
    quinoa, 206–207
    super soakers, 130–132
    thermal cookware cooking, 103
    to toast, 89
    True Skillet Cornbread, 239–240
Grana Padano cheese, in Lemon-Tarragon
    Toaster Oven Tilapia, 186–187
Gravlax, Herbed Salmon, 169–171
Greek Citrus-Honey Cake, 241–242
green beans
    Green Bean–Walnut Salad with Flaxseed
        Oil Dressing, 219–220
    One-Pot Prep: Potatoes and Green Beans,
        218–219
    to passively blanch, 72, 74
    "Teapot" or Passively Blanched Green
        Beans, 216–217
Green Restaurant Association, 154
GreenChill Advanced Refrigeration
    Partnership, 151
greenhouse gas emissions
    cows, 119, 123
    food production, transportation, and
        consumption, vii, 122
    food waste, 141, 149
    home heating and cooling, viii
greens, to grow, 153

greenwashing, 116
grills, 5, 38–42

haybox cooking
    for braises and stews, 81
    method, 101–103
    Web site for plans, 103
Hazelnut Chicken Salad on Shredded Napa
    Cabbage, 178–179
heat capacity, 8–10
heirloom produce varieties, 118
Herbed Salmon Gravlax, 169–171
Home-Style Glass Noodles, 190–191
Honey-Citrus Cake, Greek, 241–242
hormones, synthetic, 114, 120, 121
hot kitchen zones. *See* appliances; ovens
humane farming practices, 115–116

Ice
    freezer packs, 16–17, 73, 152
immersion blenders, 36
induction burners, 29–31
ingredients
    energy synergy, 129–130
    grains and pastas, 130–134
    meats, 119–126
    nuts, 134–135
    *See also specific ingredients*
Institute of Food Technologists, 141, 142–143

Johnson, Paul, 127

kitchens
    cleaning, 42–44
    cold zones, 16–19
    dry zones, 36–38
    hot zones, 25–36
    wet zones, 19–25
    *See also* appliances
Kuhn Rikon, 103

labels
    cage-free, 115, 116
    Certified Naturally Grown, 114
    country of origin, 113
    Energy Star, 18
    fair trade, 117
    Free Farmed, 115
    free range, 116
    grass-fed, 115
    greenwashing on, 116
    humane certification, 115–116
    obstacles to obtaining certification labels, 115
    organic foods, 5–6, 113–114
    sustainable fisheries, 128

## ENVIRONMENTAL BENEFITS STATEMENT

**Perseus Book Group** saved the following resources by printing the pages of this book on chlorine free paper made with 100% post-consumer waste.

| TREES | WATER | ENERGY | SOLID WASTE | GREENHOUSE GASES |
|---|---|---|---|---|
| 83 | 30,153 | 58 | 3,872 | 7,264 |
| FULLY GROWN | GALLONS | MILLION BTUs | POUNDS | POUNDS |

Calculations based on research by Environmental Defense and the Paper Task Force. Manufactured at Friesens Corporation